GRAVITY CURRENTS:
In the Environment and the Laboratory

ELLIS HORWOOD SERIES IN ENVIRONMENTAL SCIENCE

Series Editor: R. S. SCORER, Emeritus Professor and Senior Research Fellow in Mathematics and Environmental Technology, Imperial College of Science and Technology, University of London

A series concerned with nature's mechanisms — how earth and the species which inhabit it fit together into a dynamic whole, and the means by which evolution has taught them to survive.

We are *not* primarily concerned to exploit the environment to human advantage, although that may happen as a result of understanding it.

We are interested in the basic nature of the physical world, the special forms that it takes on earth, the style of life species which exploit special aspects of nature as well as the details of the environment itself.

ATMOSPHERIC DIFFUSION, 3rd Edition
F. PASQUILL and F. B. SMITH, Meteorological Office, Bracknell, Berks
LEAD IN MAN AND THE ENVIRONMENT
J.M. RATCLIFFE, Visiting Scientist, National Institute for Occupational Safety and Health, Cincinnati, Ohio
THE PHYSICAL ENVIRONMENT
B. K. RIDLEY, Department of Physics, University of Essex
CLOUD INVESTIGATION BY SATELLITE
R. S. SCORER, Imperial College of Science and Technology, University of London
GRAVITY CURRENTS: In the Environment and the Laboratory
JOHN E. SIMPSON, Department of Applied Mathematics and Theoretical Physics, University of Cambridge

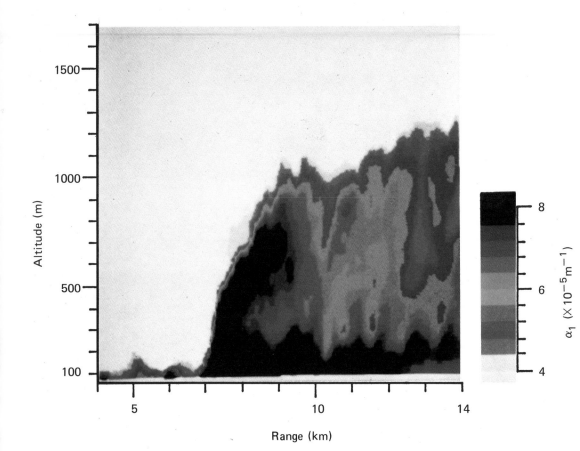

GRAVITY CURRENTS:
In the Environment and the Laboratory

JOHN E. SIMPSON, B.A., M.A., Ph.D.
Research Associate
Department of Applied Mathematics and Theoretical Physics
University of Cambridge

ELLIS HORWOOD LIMITED
Publishers · Chichester

Halsted Press: a division of
JOHN WILEY & SONS
New York · Chichester · Brisbane · Toronto

First published in 1987 by
ELLIS HORWOOD LIMITED
Market Cross House, Cooper Street,
Chichester, West Sussex, PO19 1EB, England
The publisher's colophon is reproduced from James Gillison's drawing of the ancient Market Cross, Chichester.

Distributors:

Australia and New Zealand:
JACARANDA WILEY LIMITED
GPO Box 859, Brisbane, Queensland 4001, Australia

Canada:
JOHN WILEY & SONS CANADA LIMITED
22 Worcester Road, Rexdale, Ontario, Canada

Europe and Africa:
JOHN WILEY & SONS LIMITED
Baffins Lane, Chichester, West Sussex, England

North and South America and the rest of the world:
Halsted Press: a division of
JOHN WILEY & SONS
605 Third Avenue, New York, NY 10158, USA

© 1987 J. E. Simpson/Ellis Horwood Limited

British Library Cataloguing in Publication Data
Simpson, John E.
Gravity currents: in the environment and the laboratory. —
(Ellis Horwood series in environmental science)
1. Gravitation
I. Title
531'.14   QC178

Library of Congress CIP data available

ISBN 0-85312-972-X (Ellis Horwood Limited)
ISBN 0-470-209771-1 (Halsted Press)

Phototypeset in Times by Ellis Horwood Limited
Printed in Great Britain by R. J. Acford, Chichester

Table of contents

△

Foreword

By Steve Thorpe,
Professor of Oceanography, The University, Southampton, UK

The day of the gifted amateur scientist is past. No longer may we see his enthusiasm conveyed in a text which will both draw the attention of professional scientists and stimulate the interest of students. Books by learned academics, on the other hand, are sometimes so analytical and abstract that they fail to convey that element of surprise, even awe, which first arouses a spirit of enquiry and the formulation of questions, and leads to the development of experimental results and theoretical description.

Helped perhaps by his background, particularly a lifelong hobby, John Simpson has found a nice compromise. John was a school teacher when I first met him in the late nineteen sixties and, for many years before, had been a very active glider pilot. He has flown over 1000 hours in 40 different types, and it was the effect of sea-breeze fronts on gliders which first interested him in that phenomenon. He "composed" (his own words) a movie on the subject in 1967, and was awarded the Darton Prize of the Royal Meteorological Society for work published in the journal "Weather". In 1971 he was appointed as a Research Assistant in the Department of Geophysics in the University of Reading (who must have counted themselves fortunate indeed!) and later, in 1976, he became a Research Associate in the Department of Applied Mathematics and Theoretical Physics at the University of Cambridge. There, in 1981, he gained his PhD, 45 years after the award at the same university of his BA. On a recent visit to China he surprised his hosts by asking to be allowed to see the bore on the Qiantang River (see Section 7.4 and Fig. 7.14), a rare request. His hobby had widened, and his enthusiasm shows.

In first perusing this book the reader will immediately be aware, as I was, of the number of clear and simple diagrams, the beautiful photographs which so vividly illustrate the text, and the extraordinary range of phenomena which belong to the genus "Gravity Currents". Many of them are of considerable practical importance in areas such as mining, the discharge of power station effluent or in the chemical industry, as well as in aircraft operations. Others are associated with some of the most impressive natural processes, for example sandstorms and volcanoes. John has himself contributed much to the explanation of these phenomena, especially by his careful laboratory studies. The book provides ample evidence of his academic, analytical approach to physical problems but is also a testimony to the author's delight in natural phenomena, conveying a spirit of wonder in the drama, magnificence and power of the geophysical fluids which are around us.

I welcome this book and hope that it may, in some readers at least, inspire a desire to see for themselves the phenomena it describes and to understand how they come about.

Preface

The need for this book was apparent several years ago when I prepared a short review on the subject of gravity currents. This was limited to only twenty pages and as I looked into wider environmental aspects I could see the value of a more comprehensive review, of at least ten times the length, on the manifestations of gravity currents and the closely related internal bores and solitary waves. Workers in many scientific disciplines, who are not experienced in fluid mechanics, could benefit from an account of what is already known about the properties of gravity currents and would be able to see how much is applicable to their own specialist subject. As the work developed I could see that the material described, with the many photographs and diagrams I was able to include, would also be of interest to the general reader.

The first part of the book, from Chapter 1 to Chapter 10, deals with the nature of buoyancy-driven flows in the atmosphere and oceans and on the earth's surface. These flows in the atmosphere are at scales from a cold gravity current through an open door of a house to vast squall lines of cold dense air which are hazardous to aircraft. In the ocean the study ranges from the Gulf Stream to river fronts in estuaries and fjords. The formation of these fronts is of biological importance; other important gravity currents are heated effluents from power stations. The section on earth sciences deals with snow avalanches and volcanic gravity currents.

Industrial problems in which gravity currents play an important part are described; a topical example is the spread of a cloud of dense gas which may be poisonous or explosive. The study of gravity currents of gases in mines has a long history.

The second part of the book, Chapter 11 to Chapter 17, deals with "the anatomy of a gravity current", and examines the many factors which affect their behaviour, including, as well as the nature of the head of the current, the influence of stratification and turbulence in the surroundings. The topics are dealt with by numerous laboratory experiments, and only simple mathematics is included. The final Chapter mentions briefly the rapidly growing subject of numerical models of gravity currents.

References are collected together at the end of each Chapter to avoid interrupting the text.

Many scientists kindly gave permission for use of striking photographs of phenomena in the environment; most of the laboratory photographs are my own.

I am grateful for the help of Professors R. S. Scorer and S. A. Thorpe and I am indebted to many other friends, especially the following, each of whom commented on the Chapter dealing with his speciality: Drs Tim Davies, Herbert Huppert, Jim McQuaid, Alf Mercer, Roger Smith, Steve Sparks, and Alan Thorpe. Dr Chris Bertram read the whole manuscript and suggested many improvements in the presentation. Members of the Department of Applied Mathematics and Theoretical Physics at Cambridge have helped me in many ways: Drs Paul Linden and Jim O'Donnell made many helpful comments. My thanks go to Margaret Downing for her able preparation of the diagrams.

1
The nature of gravity currents

1.1 INTRODUCTION

Gravity currents, sometimes called density currents or buoyancy currents, occur in both natural and man-made situations. These currents are primarily horizontal flows and may be generated by a density difference of only a few per cent.

An important part is played in many different scientific disciplines by gravity currents. In the atmosphere, for example, most of the severe squalls associated with thunderstorms are caused by the arrival of an enormous gravity current of cold dense air. One such advancing atmospheric gravity current in the Sudan is shown in Fig. 1.1. In this case the dense air, which is moving from right to left in the picture, is

Fig. 1.1 — The front of a gravity current of cold air in the atmosphere, made visible by suspended sand and dust.

clearly outlined by sand and dust which have been raised from the ground by the strong turbulent wind. The dust cloud is about 1000 m high and the front is advancing at about 25 m s^{-1}; some idea of the scale can be gained from the houses which can just be made out in the distance.

Knowledge of the properties of these gravity currents is obviously important for aircraft safety. The fronts produce large changes in horizontal wind and areas of intense turbulence. As they are not always so clearly marked by dust as the example in the photograph, it is possible to fly into them without any warning. Encounters of this kind have been responsible for serious accidents, at both take-off and landing.

Another, less intense, manifestation of atmospheric gravity currents appears in the sea-breeze front. These fronts form near the coast, and many of them propagate up to 200 km inland. They have important effects on the transport of airborne pollution, and also on the distribution of insect pests.

Avalanches of airborne snow, which are a severe hazard in the mountains, are gravity currents in which the density difference is supplied by the suspension of snow particles. For many years attempts have been made to reduce the damage caused by avalanches and there are research establishments solely devoted to the investigation of this special type of gravity current.

An industrial problem which has received much attention recently is the accidental release of a dense gas, which may be poisonous or explosive. Serious accidents have occurred in the resulting spread which usually starts as a gravity current. Much experimental and theoretical work has been carried out on this problem, leading to possible methods of controlling such escapes.

Even in the home, problems with gravity currents are common. If the door of a warm house is held open for a few seconds on a cold day it is easy to detect the gravity current of cold air flowing along the ground into the house.

This open door experiment is recommended to the reader who may care to use soap bubbles or puffs of smoke to detect the sudden onset of the gravity current of dense cold air after the door has been opened. This topic is dealt with in more detail in Chapter 6.

In the ocean, large volumes of warm or fresh water, less dense than the neighbouring salty water, flow as gravity currents along the surface. Gravity currents in the ocean are not so obvious to the casual observer as some atmospheric gravity currents, but lines of foam and debris on the surface may point to their presence. These lines are caused by the convergence of the flows there, and are well known to fishermen, since these currents have important effects on the distribution of fish.

Fresh-water gravity currents often flow along the surface in estuaries and fjords, above the more dense sea water. Fig. 1.2 shows an echo-sounding of such a surface flow, made in the Fraser River in Canada. This shows the cross-section of a gravity current of fresh water advancing from the right, above the denser salt water from the sea. The leading edge of this current has a "head" which is deeper than the following flow, a feature which is seen in most gravity currents.

The oil slick is an example of a man-made environmental problem. An oil spillage from a ship forms a non-mixing gravity current of less dense fluid on the sea surface. It is important to understand the development of this flow and find possible methods for both its containment and its dispersal.

Fig. 1.2 — An echo-sounding made in the Fraser River in Canada. The front of a gravity current of fresh water is moving above sea water. (Courtesy of David Farmer).

1.2 DAM BREAK

To understand the physics of a gravity current it will help to consider what happens when the wall of a dam breaks and releases the water behind it.

A mass of suddenly released water will start to collapse and flow horizontally. The main force acting on the water in such a flow is due to gravity, and acts vertically downwards. This results in a downward motion of the water which can only occur if the water spreads horizontally. So the potential energy of the water due to its height is continuously converted into the kinetic energy of the horizontal motion.

If the flow spreads mainly in one direction, for example along the bottom of a valley, a rough idea of the velocity, U, in that direction can be obtained by equating the values of the potential energy loss and the kinetic energy gain, i.e.

$$(mU^2)/2 = mgH/2$$

or

$$U = \sqrt{gH}$$

where m is the mass, $H/2$ the mean height of the centre of gravity and g the acceleration due to gravity. If for example the water was originally 20 m deep, the velocity would be about 14 m s^{-1}, or roughly 30 m.p.h.

Viscous forces in the fluid can also have important effects. Viscosity may be likened to friction in that a viscous fluid exerts retarding forces on those parts of itself which are trying to move with greater velocity than the rest, just as retarding forces due to friction occur between two solid surfaces in relative motion. The lower layers of the water in the dam-break flow are retarded by the ground, and have a considerable effect on the form of the leading edge of the fluid.

1.3 GRAVITY CURRENTS

The water in a dam-break flow is submerged in the atmosphere, but this has only a
very small effect on its behaviour. If the air is replaced by a fluid which is only a few
per cent less dense than the collapsing fluid, then the flow will be different.

If in this "dam-break analogy" flow the density difference between the two fluids
were only 1 per cent, the effective driving force would then have been reduced to
only 1 per cent of normal. Unless the coefficient of viscosity is large, or the scale is
very small, the main controlling forces will be gravitational and inertial, due to the
displacement of the fluid around the advancing current. Due to the net gravitational
acceleration of the collapsing fluid being now only g'/100, the previous dam-break
flow will be replaced by one appearing to move "in slow motion".

A typical *gravity current* of dense fluid is now moving forwards into a slightly less
dense fluid. In this case its rate of advance U can be approximated by

$$U = (gH/100)^{\frac{1}{2}}$$

or, in general, if ρ is the density of the less dense fluid and $\Delta\rho$ is the density difference,

$$U = \left(\frac{\Delta\rho}{\rho} gH\right)^{\frac{1}{2}}$$

The term $g(\Delta\rho/\rho)$ will usually be denoted by the symbol g', and called "reduced
gravity".

The fluid in a gravity current may be chemically different from the surroundings
and have a different molecular weight, but often the difference in specific weight that
provides the driving force is due to dissolved material or to temperature differences.
The large-scale gravity current in the atmosphere shown in Fig. 1.1 was caused by
temperature differences. If the temperature difference was about 12°C, this would
give a density difference of about 4%. The value of g' will be 0.39 m s^{-2} With a
current height of 1000 m, we would expect the rate of advance to be about $(g'H)^{\frac{1}{2}}$
which is just under 20 m s^{-1}.

1.3.1 Suspension flows; turbidity currents

One way in which the overall density of a fluid can be increased is by the suspension
of many small dense particles within it. Such *suspension currents* may be formed in
various ways. One of the most important processes is the raising of material from the
ground and its suspension by the turbulence within a gravity current. This suspended
material increases the density and hence the speed and turbulence within the current.
The process can thus become "self stoking" in a current on a slope, further increasing
the strength of the current. An example of such a suspension current is shown in Fig.
1.3. This illustrates a suspension current of kaolin in water advancing through water
towards the camera in the laboratory. This photograph shows very clearly the

Fig. 1.3 — A suspension current in the laboratory, advancing towards the observer. (Courtesy
of J. R. H. Allen).

complicated shifting instability patterns manifested by a gravity current advancing
along a plane surface.

Self-stoking gravity currents containing suspended matter also occur in the
ocean. They start on slopes near the coast as mud-slides which increase in intensity
until a suspension current is formed. These *turbidity currents* may become large
enough to travel at speeds of over $30 \, m \, s^{-1}$. They can gouge out vast channels in the
sea bed and their progress has been followed by monitoring the successive breaking
of submarine telephone cables, showing that they can travel for hundreds of
kilometres.

1.4 BORES

A related phenomenon is the **bore**, which is also concerned with mass transport and
has many features in common with the gravity currents already described.

The best-known type of bore is a tidal disturbance which moves upstream in some
rivers and may be very violent at spring tides. It is an example of a *hydraulic jump* in
which there is a sudden increase of the water depth associated with a change in the
flow rate.

If the increase of depth at the front of a bore is less than about a third of the
undisturbed water depth, a series of smooth waves appears at the leading edge and
the bore is called "undular". For larger steps the tidal bore is turbulent and it
advances as a wall of tumbling breakers; the structure is similar to that of breakers
advancing towards the sea-shore.

Bores appear in rivers where the tidal range is large and the form of the estuary is suitable. The conditions are favourable in several rivers in England; on the River Severn the bore has been ridden by experienced surfers and journeys of two or three miles have been achieved. However, since they are produced by tides, surfers who miss their moment may have to wait over 12 hours for the next breaker! Fig. 1.4

Fig. 1.4 — A surfer on the advancing bore on the River Severn. (Courtesy of the Severn–Trent Water Authority).

shows a rider on a surf board using one of the waves at the front of the Severn bore. The bore shown here is an undular one, but breaking waves are forming in the shallower water near the banks of the river.

The conditions ahead of a hydraulic jump or a bore can be related to those behind it in a simple mathematical treatment.

In a moving frame of reference the jump is brought to rest as in Fig. 1.5 where h_0, U_0 and h_1, U_1 are the height and velocity on the two sides of the jump.

The volume flux Q per unit width is

$$Q = U_0 h_0 = U_1 h_1$$

If density is ρ, since mean pressures at both sections are

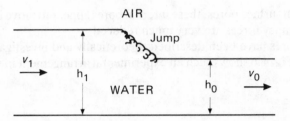

Fig. 1.5 — The flow through a hydraulic jump. The axes have been taken to bring the jump to rest, and h_1, U_1 and h_0, U_0 are the height and the velocity on the two sides of the jump.

$$(g\rho h_0)/2$$

and

$$g(h_0-h_1)/2$$

the equation of momentum is

$$Q(U_1-U_0)=g(h_0^2-h_1^2)/2$$

hence

$$Q^2=gh_0h_1(h_1+h_0)/2$$

The *energy*, however, does not balance and the loss of energy per unit time is

$$\rho(U_1^2-U_1^2)/2+g\rho(h_0-h_1)$$

or

$$g\rho(h_1-h_0)^3/4h_1h_0$$

This loss of energy, which must occur at a bore, is mostly effected in the undular case by the waves, each of which carries energy as it moves away from the front. The more intense bores cannot carry away enough energy by this method and the energy excess is dissipated by turbulence in the tumbling breakers at the leading edge.

1.5 INTERNAL BORES

The previous section considered bores at the free surface of a water flow. A somewhat similar class of *internal bores* can be formed at an interface between two fluids, one lying on top of another which is perhaps only a few per cent denser.

Compared with surface bores, these internal bores appear to move in "slow motion", since the buoyancy forces are very much reduced.

Internal bores have been described theoretically and investigated in laboratory experiments. Fig. 1.6 shows such an experimental arrangement in which an obstacle

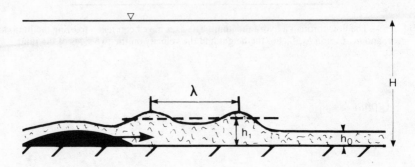

Fig. 1.6 — The production of an internal bore in an experimental laboratory tank, in which a layer of fresh water lies above salt water. The bore is moving to the right along the interface between the two fluids, faster than the moving obstacle.

is towed along the bottom of a tank containing a two-fluid system. The obstacle is moving to the right and displacing an undular internal bore which steadily moves along the interface ahead of it. Such experiments have been very helpful in understanding the physics of undular bores and will be described in more detail in Chapter 13.

During the last few years internal bores have provided an explanation for an increasing number of phenomena in the environment, both in the ocean and in the atmosphere. They may for example be formed in the ocean by tidal effects on fresh-water layers near the coast. In the atmosphere they are formed in dense stable layers by advancing flows of cold dense air from thunderstorms, and they are also associated with sea-breeze fronts.

When the ratio of the height h_1 behind the jump (see Fig. 1.6) to that in front of it, h_0, is less than about 2, then the internal bore is undular. In much deeper internal bores the leading edge is turbulent and appears very similar to the front of a gravity current. The photograph in Fig. 1.7 shows the clouds forming at an atmospheric undular bore in northern Australia. This phenomenon appears in the early morning and is marked by a spectacular roll of cloud; its striking appearance has led to its name, the "Morning Glory".

1.6 SOLITARY WAVES

"Gravity currents" and "gravity waves" are sometimes confused and the distinction between them is not always apparent. Here the name "gravity current" will be applied to a phenomenon in which there is a clear transfer of mass (usually

Fig. 1.7—Clouds marking an undular internal bore in the atmosphere, the "Morning Glory" in northern Australia. (Courtesy of Roger Smith).

horizontal). In gravity waves there is little transfer of mass and the main transport is that of energy.

An undular bore, as has been noted, consists of an increase in depth of a fluid advancing with a series of waves on its surface. Closely related is the "solitary wave", which is another shallow-water phenomenon, i.e. a disturbance which is high compared with the undisturbed depth. A solitary wave is not a periodic wave but consists of a single symmetrical hump which propagates at uniform velocity without change of form.

Scott Russell in the 1840's investigated solitary waves on the surface of a canal. On horseback he was able to follow examples of such waves for several kilometres and he showed that, with length, depth and amplitude properly matched, a solitary wave can propagate virtually unchanged, except for small effects due to bottom friction which reduce the size of the wave.

Internal solitary waves can exist at an interface between two fluids of different density. Examples occur in the atmosphere where solitary waves have been observed on stable layers, moving steadily away from the distant disturbances which generated them. What is observed on the ground is somewhat similar to the arrival of a bore, with a line of cloud and a gust of wind, but in this case with only a temporary increase of surface pressure.

In the laboratory these internal solitary waves are very easy to create. Fig. 1.8 shows an example of an internal solitary wave formed at a layer which had been laid down by a gravity current. When the gravity current reached the end of the tank

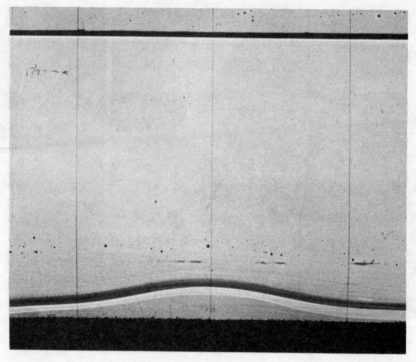

Fig. 1.8 — Internal solitary wave moving along an interface between two fluids in a laboratory tank.

some of the dense fluid ran up the wall and then descended as a mass which generated the solitary wave shown, moving from right to left. When this disturbance reached the other end of the tank, 3 m distant, it was reflected and returned with nearly the same shape and speed.

2

Atmospheric gravity currents

Gravity current fronts and internal bores have many manifestations in the atmosphere, some at a very large scale. A number of case studies of atmospheric observations will be examined to see how far they can be explained in terms of the physical processes described in Chapter 1.

Experiments with dense saline flows in water tanks have elucidated features of dense atmospheric flows. This is because many of the results of buoyancy in a compressible atmosphere can be deduced from an incompressible fluid if **potential temperature** is substituted for density (with the difference in sign noted). The potential temperature of the air at any point is the temperature this air would attain if its pressure was changed to 1000 mb (standardised ground level), with no gain or loss of heat.

A zone of constant potential temperature in the compressible atmosphere can be modelled in the laboratory by a salt solution of constant density. A tank of salt solution with its concentration and density decreasing with height is stable and corresponds to a stable atmospheric zone with potential temperature increasing with height.

The *dynamic similarity* between the gravity currents seen in the laboratory and those seen in thunderstorm outflows has been established in many case studies [1,2]. The requirements for formal similarity are (A) geometric similarity between the model and the large scale flow, and (B) the equality of the relevant dimensionless numbers. In this context the two most important dimensionless numbers are the *Reynolds number*, Re, and the *Froude number*, Fr.

The Reynolds number is a dimensionless ratio which gives a criterion for the critical behaviour of a flow , depending on its speed, depth and coefficient of viscosity. The Reynolds number of a flow at velocity U, depth h and kinematic viscosity v is given by Uh/v. Reynolds showed, by using a slender trail of dye through the centre of a pipe, that the flow would change from a laminar to a turbulent one for Reynolds numbers greater than about 600.

As will be seen in the descriptions of gravity current experiments, when the Reynolds number is greater than about 1000, the flow patterns are independent of its value. As the Reynolds number in most thunderstorm outflows is of the order of 10^8, this ratio is unimportant in determining the nature of the flow.

The most important dimensionless number related to the flow of atmospheric gravity currents is the internal Froude number, the ratio of inertial forces to buoyancy forces, usually written in the form $\mathrm{Fr} = U/(g'h)^{\frac{1}{2}}$. The velocity of the current is U, g' is reduced gravity and h is the depth where the density discontinuity occurs. The Froude number of cold outflows when they have been measured is not very different from 1, in agreement with the smaller-scale laboratory results.

Laboratory experiments have been able to include a number of conditions of the ambient fluid, such as uniform velocity with a simple shear profile, and density profiles with a step and with uniform variation. It has not been possible to model all the complicated atmospheric configurations of wind and density profiles, and other sources of information may be able to supplement laboratory results. One important source is the numerical model, as described in Chapter 18.

2.1 THUNDERSTORM OUTFLOWS

Thunderstorms are generated by warm moist air rising in unstable conditions. Eventually this rising air reaches the boundary formed by a very stable layer called the tropopause. Here it spreads out, forming the familiar "anvil" cloud. During the early stages of anvil development, the ice-crystal cloud sometimes gives a good indication of the form of the flow patterns. An example of this is shown in Fig. 2.1

Fig. 2.1 — Clouds at an active thunderstorm, showing developing anvil cloud spreading to the right above the storm. At 30 000 feet above Turin, on 12 August 1972. (Courtesy of Colin Street).

which was taken from an aeroplane just beneath a developing anvil cloud at about 30 000 ft, above Turin in Italy. This anvil cloud can be seen to have the form of a gravity current moving towards the right beneath a rigid boundary. It has the typical contour of the front, but the bulges visible beneath it are probably due to fall-out of ice crystals which often forms "mamma" clouds.

Rain and hail, falling in another part of the thunderstorm, produce a down-draught of cold air that descends towards the ground. When the cloud-base is high, the rain falls an appreciable distance through this non-saturated zone and the cooling due to evaporation may be very great. The cold column of air, which may be several per cent denser than the surrounding air, reaches the ground and spreads out horizontally, forming a gravity current.

A schematic diagram of the structure of a thunderstorm cell is shown in Fig. 2.2.

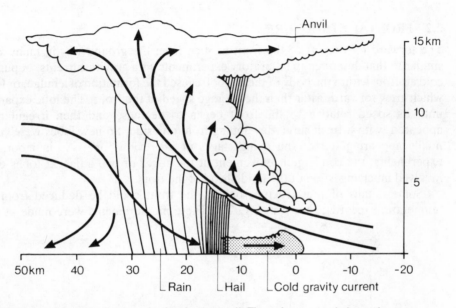

Fig. 2.2 — Schematic diagram of a thunderstorm cell. Flows shown are relative to the storm, which is travelling to the right.

The downdraught which spreads out away from the storm is shown here as falling behind the area of rising currents, but in fact moves perpendicular to the plane of the paper, as this really needs to be demonstrated in three dimensions. The downdraught section of a thunderstorm plays an important part in the structure and regeneration of severe storms, since the dense cold air pushes up the warmer and lighter air it meets. Thus it reinforces the warm updraught that creates new rain clouds in a system of thunderstorm cells.

2.1.1 Outflow observations from the ground

When a thunderstorm outflow advances over sandy and dusty land in an arid country, the strong turbulent wind following the gust front raises dust from the ground. One of the most awe-inspiring sights in nature then appears, a vast wall of dust 1000 metres high advancing across the country. The appearance of such a dust-laden cold outflow was shown in Chapter 1. These fronts of cold dense air may move at 20 m s^{-1}, with a wind behind the first squall gusting up to twice this speed.

These dust storms occur regularly in Arizona, USA, where their approach to the town of Phoenix has been described [1]. They appear in India, where they have been given the name of "andhi". They are also common in many parts of Africa, especially in the Sudan where they are named "haboob" [2]. (This is merely the Arabic name for strong wind.)

Haboobs are particularly prevalent near Khartoum, where they are most common in June. A haboob approaching the airfield at Khartoum in the evening of 17 June 1969 was studied in detail and a time-lapse movie was made to observe changes in the frontal structure. Measurements were taken of the wind and temperature changes, and the dust content was also recorded [3].

2.2 FRONTAL STRUCTURE

The forward movement of the front as seen from the ground was in many ways similar to that described in laboratory experiments with gravity currents. A push of cold air from within the body of the haboob caused the formation of a bulge, or lobe, which grew forward faster than the average speed of the front. The lobe expanded until its speed relative to the front began to decrease and then irregularities appeared, with a fresh surge developing from one side. So new lobes were continually appearing out of the dying stages of a previous one. As in laboratory experiments, the convergent movement at the edges of two adjacent lobes often involved much ingestion of the fluid ahead of the front.

Some details of the structure behind the front could be deduced from the temperature records, shown in Fig. 2.3. These measurements were made at roof

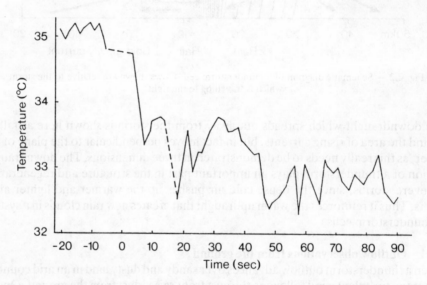

Fig. 2.3 — Temperature record obtained during passage of haboob front at Khartoum, 17 June 1969. The time is in seconds, relative to the arrival of the front. (After Lawson, 1971 [3]).

height every 1.6 seconds. They show an initial temperature drop of between 1.5° and 2°C, occurring over a distance of less than 50 metres. After this rapid drop, the mean level decreased only slowly, falling another half degree in ten minutes.

2.2.1 Dust Content
Samples of the airborne dust were collected in visibilities ranging from 100 m to 3 km by passing a known volume of air through filters and weighing it. The increase in weight never exceeded 0.1 g, the minimum detectable difference. The resultant upper concentration limit was 40 mg m^{-3}, equivalent to a density difference about 1/300 of that due to the temperature drop between the warm and cold air, so it appeared that the dust content could be neglected as a contributing factor to haboob dynamics. Similar results for dust content have been obtained in measurements made for summer dust storms in Russia, in the Mangyshlak Peninsula [4].

2.2.2 Pressure changes at surface
After the particularly dry summer of 1983 in south-east Australia, several dust squalls were observed in the Melbourne area. Fig. 2.4 shows the front of one of these

Fig. 2.4 — Haboob crossing the city of Melbourne, 2 February 1983. (Courtesy of Melbourne Weather Bureau).

"haboobs" crossing the city, where the buildings help to give an idea of the scale of the phenomenon.

A particularly clear record of one of them was obtained [5] as it passed

Aspendale, Victoria and is shown in Fig. 2.5. On 8 February at 1500 h the wind shifted from about 5 knots North to strong Southerly. The front of the dust storm advanced at 20 m s^{-1} and the speed of the initial gust was usually greater than this. The temperature fell from 35° to 28°C at the squall, and the pressure rose suddenly. Assuming that the pressure change was hydrostatic and due to the arrival of cold air through a depth h, then the depth of the dense air can be calculated. This gave an initial head height of 320 m, gradually increasing up to 1000 m within one hour.

If we interpret the flow as a gravity current, the velocity of the front is then given by

$$U = k(g\Delta T.h/T)^{\frac{1}{2}}$$

If we take $k = 1$, this gives a velocity of 22 m s^{-1}, very close to the observed value of 20 m s^{-1}.

The detection of these pressure changes at the front of a thunderstorm outflow, using an array of pressure sensors on the ground, has been used to trace its progress across country.

2.2.3 Measurements from instrumented towers

The depths of cold outflows from thunderstorms vary from only a few hundred metres to well over a kilometre and much useful information on their structure has been obtained from measurements from instrumented towers [6]. Television towers have often been used but some research organisations have their own fully equipped instrumented towers. From one such multi-level tower 461 m high in central Oklahoma [7] measurements of wind and temperature have been taken during the passage of many gust fronts and profiles of airflow and of temperature have been constructed.

A model of a typical outflow has been built up from these measurements and is shown in Fig. 2.6. The raised head and the internal flow show a similar structure to that seen in the laboratory gravity currents we will describe in later chapters. In 17 cases examined, the height of the foremost point, or nose, was around 100 m. It was found that low-level stratification affected the slope of the gust front, and that the outflow could sometimes override a dense layer near the ground, as in the intrusive flows described in Chapter 13.

In several examples of gust fronts there were secondary or even multiple surges, whose behaviour did not fit into the simple gravity current picture. Some of these appear to be related to the generation of an atmospheric bore by the gravity current.

2.2.4 Remote sensing: sonar

Although tower data provide much useful information on cold outflows, they suffer from a number of limitations. By their nature, towers can only reach a few hundred metres and outflows may be three or four times the height sampled. There is also a certain amount of luck in having a non-portable tower in the right place to catch an outflow.

Although echo sounding in the ocean dates from the 1920's, it was not until 1968

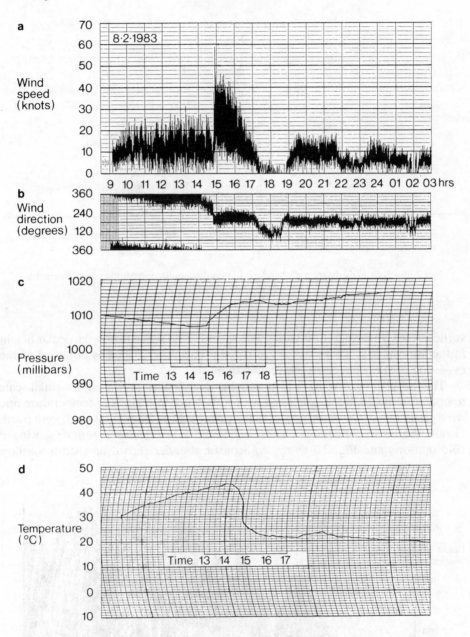

Fig. 2.5 — Surface records of passage of haboob at Aspendale, Victoria, 8 February 1983. (a)
Wind speed. (b) Wind direction. (c) Atmospheric pressure. (d) Temperature. (Courtesy of J.
R. Garratt).

that effective echo sounding was achieved in the atmosphere [8]. Improved antennae
soon permitted operation at about 1000 Hz giving a range of up to 1.5 km.

Each pulse of "acoustic radar" or *sonar* produces on the facsimile recorder a

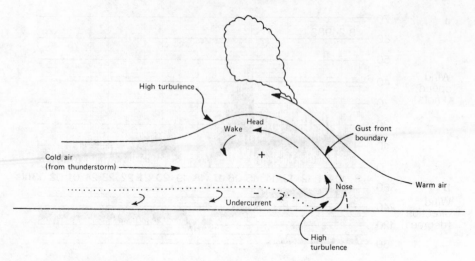

Fig. 2.6 — Schematic diagram of thunderstorm outflow, built up from tower measurements. (After Goff, 1976 [7]).

vertical trace whose darkness varies with the received signal from a particular height in the atmosphere. These traces are closely spaced, so a trace is obtained of the evolution of atmospheric structure with time.

The acoustic scattering which can be detected is produced by small-scale temperature fluctuations associated with turbulence in regions of temperature and wind gradients, so we would expect to see strong echoes from a gravity current front. Ground-based acoustic sounding has been found to be effective in remote sensing of cold outflows and Fig. 2.7 shows an acoustic sounder record of a storm outflow

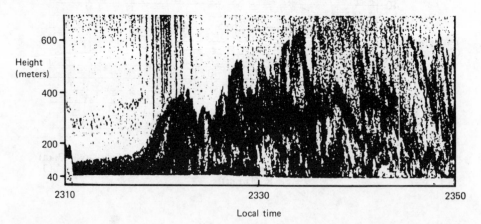

Fig. 2.7 — Acoustic sounder record of thunderstorm outflow passing Haswell, Colorado. (Reproduced with permission from McAllister, 1968 [8] copyright 1968, Pergamon Books Limited).

passing Haswell, Colorado [8]. The vertical noise-lines in the upper part of the record at 2320 are caused by wind noise associated with the passage of the front. Otherwise the echoes show clearly the boundaries of the turbulent outflow, and also give some information about the internal structure of the gravity current.

2.2.5 Remote sensing: radar

Radar has been used widely in building a model of the dynamics of the front of a cold outflow. Doppler radar, in which the change in frequency from moving targets is used, enables wind speeds to be measured at large heights and distances. The results from many tower measurements combined with radar information confirm the expectation that most atmospheric cold outflows are gravity currents with similar characteristics to those investigated in the laboratory. A narrow cold frontal band, accompanied by strong winds, tornadoes and pressure jumps, was examined by triple doppler radar as it passed through the Central Valley of California [9]. The isometric plot shown in Fig. 2.8 reveals a "classic gust front" surface.

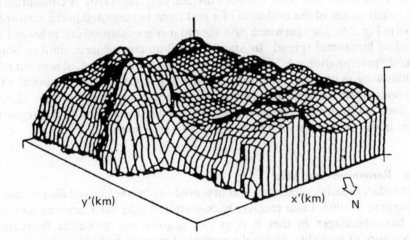

Fig. 2.8 — Perspective view of gust front, measured by doppler radar. (Courtesy of R. E. Carbone).

Using the Froude number relationship for a gravity current, the expected velocity for this front has been calculated. A more exact form, including the pressure term, gives

$$U = [\text{Fr } g\Delta z \ \{(P_2/P_1)T_1 - T_2\}/T_2]^{\frac{1}{2}}$$

where P_1, T_1 and P_2, T_2 are pressure and temperature before and after the passage of the front.

The pressure increase from numerous microbarographs recorded a surface pressure increase of 3 mb, giving results of $U = 21.4 \text{ m s}^{-1}$ and 18.3 m s^{-1} for values of $Fr = 2^{\frac{1}{2}}$ and 1.1 respectively. The observed speed of the gust front was 21.7 m s^{-1}, suggesting that the gravity current hypothesis is a likely explanation of the storm motion.

The lobe and cleft structure at a haboob has already been described, appearing similar to those seen in laboratory experiments. In the overall view of Fig. 2.8 some lobe structure is also apparent at the leading edge. Measurements of precipitation behind mesoscale cold fronts show regions of precipitation core, divided by gap regions. It has been suggested that these oriented, ellipsoidal precipitation cores and gap regions are similar to the bulges and clefts and could be due to an instability produced by the strong horizontal shear of the wind across the front.

The structure of most of the atmospheric fronts described has either been based on an almost instantaneous "snapshot" of a front, or built up from a tower as the front passed by, assuming the structure did not change during the time of passage. Particularly in the early stages of development such observations may not give a true picture.

In a doppler radar study, allowing determination of the entire vertical structure, the life cycle of the gust front has been divided into stages [10]. A conceptual model of the early stages of the evolution of a gust front as measured in the atmosphere is shown in Fig. 2.9. The rain which gave the radar echoes showed first in Stage 1 a short period of horizontal spread. In Stage 2 a roll-up commenced, until in Stage 3 a marked "precipitation roll" had appeared, and moved forward, almost cut off from the following dense flow. The form of this flow appears to be identical with the behaviour of the current in a laboratory experiment pictured in Fig. 12.8, to be described in Chapter 12. Here it is sufficient to note that a rapidly changing structure occurs in the early stages of a developing divergent flow.

2.2.6   Remote sensing: lidar

Laser radar, usually called lidar, was first used to observe aerosol distribution in the atmosphere in 1963. Lidar receives back-scattered light from airborne particulates, and has advantages in that it does not require any turbulent fluctuations in temperature or humidity, as usually employed in sonar and radar observations.

Lidar has been used in the atmosphere [11] employing the plan position indication (PPI) and range height indication (RHI) modes. Both the structure of the convective mixed layer and fronts of gravity currents have been studied. The RHI display of one of the latter is shown in the Frontispiece and will be described in more detail in Chapter 4.

2.3   DATA FROM SATELLITES

Thunderstorm outflows may travel 50 km or more from their source. They are often marked by a line of "arc" clouds, so named from their appearance from the ground as an approaching arc of cloud right across the sky. These have often been followed on radar as they move further from the generating storm, and on some occasions they have been traced by satellite imagery.

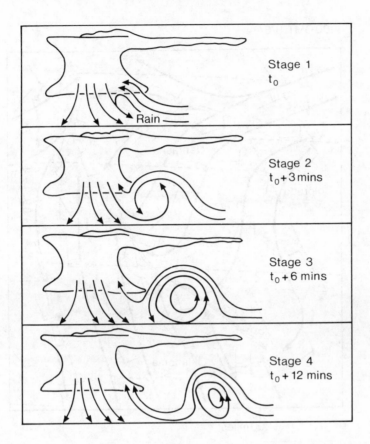

Fig. 2.9 — A model of the early stages of the evolution of a gust front, from doppler radar measurements. (After R. M. Wakimoto).

2.3.1 "Rope clouds"

The NOAA/GOES-5 satellite imagery at visible wavelength of 9 June 1984 gave a good view of a thin trail of enhanced cumulus cloud extending over several hundred miles across Kansas and Oklahoma. This example of a *rope cloud* is shown on the map in Fig. 2.10. The air flow showed that the rope line was situated at the frontal wind shift along the leading edge of a cold frontal system [12]. The progress of this line was followed for 2 hours, during which cirrus cloud masked the northern section of the line, but the rope cloud was still visible in the southern section of the 500 km long squall line.

There were no detailed surface observations of surface temperature and wind on this occasion, but records from a similar event [12] are shown in Fig. 2.11. These records of potential temperature and wind speed were made over a period of just 2 minutes from the 200 m instrumented tower of Boulder Atmospheric Observatory. The passage occurred in 10 seconds at each tower level, which converts to a frontal width of 170 m, given the 17 m s^{-1} speed of the front. The potential temperature

1900 GMT 9/6/84

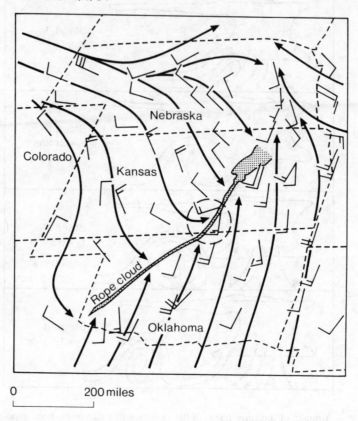

Fig. 2.10 — A rope cloud lying across Kansas and Oklahoma, NOAA/GOES-5 visible imagery.
The position of the rope line is at the frontal wind shift. (After M. A. Shapiro).

dropped by 6°C in 10 seconds during the frontal passage, and there was a frontal "nose" at about 75 m due to surface frictional drag. The substantial wind increase coincided with the temperature drop.

Numerous measurements of this kind of sharp frontal passage ahead of major frontal systems have given rise to the suggested mechanism [13] shown in Fig. 2.12. A hydraulic jump mechanism forms a sharp disturbance moving ahead of the obstacle created by the main storm. This hydraulic mechanism may consist of either a gravity current or an atmospheric bore, and the latter will be considered in Chapter 3.

2.3.2 "Global gravity currents"

Satellite photographs have detected from time to time enormous expanding circular rings of cloud , on a scale of 1000 km or more. These expanding rings have sometimes been followed for as long as one or two days. They are usually comparatively low-level clouds, outlining an area at the edge of a dense gravity current, but the origin of the dense air may have several different causes, two of which will be examined.

The first example, illustrated in Fig. 2.13 (a), was taken in the infrared from the

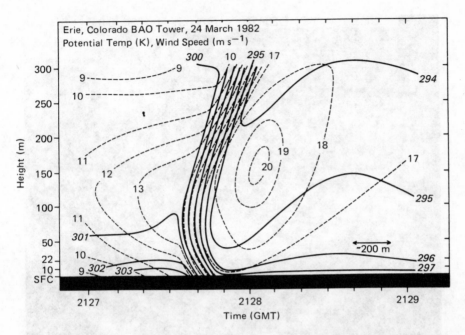

Fig. 2.11 — Records from a tower of potential temperature (K, solid lines) and front-normal wind component, (m s^{-1}, dashed lines). (Courtesy of M. A. Shapiro).

Fig. 2.12 — Suggested mechanism responsible for rope cloud observed on 4 May 1983. (Courtesy of K. L. Seitter).

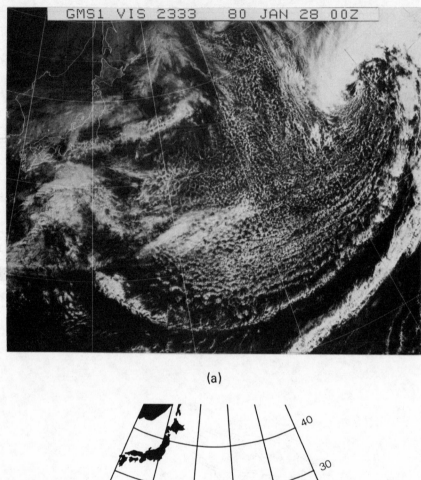

(a)

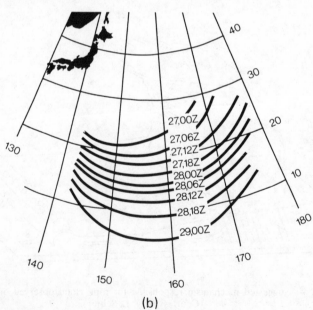

(b)

Fig. 2.13 — Sharp line covering 30 degrees of latitude moving from the south-east of Japan. (a) From Japanese Geostationary Satellite, infrared image. (b) Traces at 6 hour intervals, showing progress at about 8.3 m s^{-1}. (Courtesy of R. Kimura).

Japanese Geostationary Meteorological Satellite (G.M.S.) on 27–29 January 1980. A depression over the Pacific, south-east of Japan, contains a sharp line which extends across 30 degrees of latitude. This front was followed by air coming from Siberia which was 10°C colder than the sea surface and, during the consequent heating, instability patterns can be seen for hundreds of kilometres behind the front. Fig. 2.13 (b) gives traces at 6 hour intervals, showing progress towards the south of about 30 km in each hour, or an average speed of 8.3 m s^{-1}.

The second example examines a wind called the "Tehuantepecer". Ships traversing the Gulf of Tehuantepec on their way to and from the Panama Canal have often encountered a special kind of squall line moving from the north. An arc of cloud approaches across the sky from horizon to horizon, bringing with it a strong squall of cold air, quickly followed by clearing skies. The map shown in Fig. 2.14 displays the

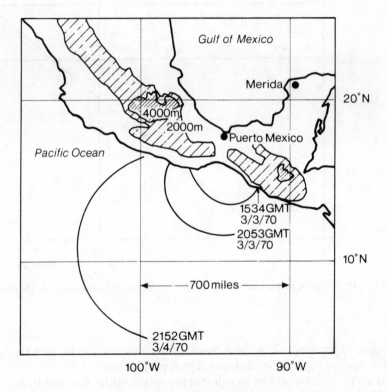

Fig. 2.14 — The progress of a "Tehuantepecer" in March 1970 as it moved out across the Pacific Ocean. (After Parmenter, 1970 [14]).

progress of such a "Tehuantepecer", drawn from a series of photographs taken by satellite ESSA 9 on 3 February 1970. The position of the squall line is shown as it moved from the south coast of Mexico through the Gulf of Tehuantepec into the Pacific.

It has been shown that these winds are a result of both the prevailing weather situation and the topography [14]. In the winter, cold continental outbreaks often penetrate south into the Gulf of Mexico and the front of this cold air advances until it reaches the range of mountains across Mexico. These mountains form a range 4000 metres high, but there is a mountain pass almost down to sea level dividing them, north of the Gulf of Tehuantepec. When the cold air reaches this point it overflows through this natural spillway and rushes in a torrent through the opposite slope across the open gulf to the southward.

When the first satellite photograph was taken of the cloud line, the front had already been travelling for about 5 hours into the Pacific, and the map shows three subsequent positions of the advancing gravity current. During 24 hours the line travelled a distance of 1000 miles, and Fig. 2.15 shows a log–log plot of the position

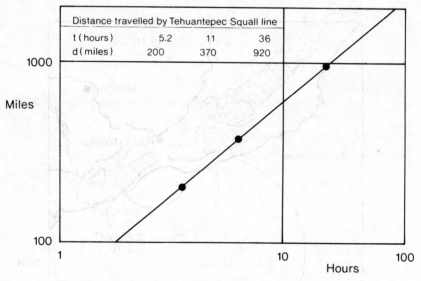

Distance travelled by Tehuantepec Squall line			
t (hours)	5.2	11	36
d (miles)	200	370	920

Fig. 2.15 — Log-log plot of distance moved by Tehuantepecer during 36 hours. The slope of the line is 0.8.

with time of the front. The three points plotted in this way lie on a straight line, of slope 0.8, showing that the distance $R(t)$ is proportional to $t^{0.8}$.

In Section 12.5 it will be shown that the relationship obtained from dimensional analysis and confirmed by laboratory experiments is

$$R(t) = k(Q\ g')^{\frac{1}{3}}\ t^{\frac{2}{3}}$$

where Q is the mass flow per unit width of the dense flow at the source. We have no information about the changes in Q, but it is reasonable to suppose that Q continued to increase after the initial arrival of dense air at the pass. If a uniform increase with time is conjectured then $R(t)$ will vary with $t^{\frac{4}{5}}$, very close to the observed figures.

It is of interest to examine to what extent this type of large-scale gravity current, flowing throughout a period of a day or longer, will be affected by the earth's rotation. Chapter 17 will discuss the effects of Coriolis forces on a gravity current advancing under various conditions. The relevant quantity is the Rossby radius, L, a measure of distance of travel before the earth's rotation produces significant effects. This depends on the ratio of the velocity, given in the form $(g'h)^{\frac{1}{2}}$ and the Coriolis parameter, $f = 2\,\Omega\sin\theta$, where Ω is the angular velocity of the Earth and θ is the latitude. The relationship can be written

$$L = \{(g'H)^{\frac{1}{2}}\}/f$$

The three cases of "global gravity currents" we have described are all situated in latitude 10° to 20° (either N or S). This is not very far from the equator, so the Coriolis force will be small, leading to a large value for the Rossby radius. Taking approximate values of $f = 2\text{x}10^{-5}$, $g' = 0.1$ m s^{-2} and $H = 500$ m, a value for L of 350 km is obtained. (This is in the atmospheric boundary layer, in which we expect the effect of rotation to be very much further reduced.)

BIBLIOGRAPHY

[1] Idso, S. B. *et al.* 1972. American haboob. *Bull. Amer. Met. Soc.*, 53: 930–935.

[2] Sutton, L. J. 1951. Haboobs. *Quart. J. R. Met. Soc.*, 57: 143–161.

[3] Lawson, T. J. 1971. Haboob structure at Khartoum. *Weather*, 26: 110–112.

[4] Kharitanova, S. S. 1969. Dust content of the air during dust storms in the Mangyshlak Peninsula. *Met. i. Gidro.*, No.5, 87–88.

[5] Garratt, J. R. 1984. Cold front and dust storms during the Australian summer, 1982-3. *Weather*, 39: 98–103.

[6] Hall, F. F., Neff, W. D. & Frazier, T. V. 1976. Wind shear observations in thunderstorm density currents. *Nature,* 264: 408–411.

[7] Goff, R. C. 1976. Vertical structure of thunderstorm outflows. *Mon. Wea. Rev.*, 104: 1429–1440.

[8] McAllister, L. G. 1968. Acoustic sounding of the lower troposphere. *J. Atmos. Terr. Phys.*, 30: 1439–40.

[9] Carbone, R. E. 1982. A severe frontal rain band. Part 1. *J. Atmos. Sci.*, 39: 258–279.

[10] Wakimoto, R. M. 1982. The life cycle of thunderstorm gust fronts. *Mon. Wea. Rev.*, 110: 1060–82.

[11] Shimizu, H. *et al.* 1985. Large scale laser radar for measuring aerosol distribution over a wide range. *Applied Optics*, 24: 617–626.

[12] Shapiro, M. A. 1984. Meteorological tower measurements of a surface cold front. *Mon. Wea. Rev.*, 112: 1634–9.

[13] Seitter, K. L. & Muench, H. S. 1985. Observation of a cold front with rope cloud. *Mon. Wea. Rev.* 113: 840–848.

[14] Parmenter, F. C. 1970. A Tehuantepecer. *Mon. Wea. Rev.* 98: 479.

3

Atmospheric bores

A variety of atmospheric observations, especially during the last ten years, have shown what seem to be internal bores propagating on stable layers such as temperature inversions. An abrupt increase in ground level pressure (several millibars in a few minutes) is followed by a sustained period of high pressure often containing wavelike oscillations. The increase in ground level pressure is often accompanied by an increase in ground level temperature and a shift in wind in the direction the disturbance is moving. There may be distinctive cloud formations, sometimes a roll cloud, perhaps followed by a series of rolls. These disturbances can travel several hundred kilometres from their point of generation. A study of undular pressure lines has been made in the arid region of northern Australia [1], and their development has been traced over long distances.

Some early observations of these pressure jump lines were made in the Midwest United States [2], and it was speculated that these disturbances were generated by an impulsive motion of a cold front into an existing nocturnal inversion, producing a propagating bore or solitary wave. As will be seen in Chapter 13, it is possible for the steady advance of a front of cold fluid to generate an internal bore in a stable layer over a wide range of depths and density differences.

Similar case studies made in North America [3] show that the cold outflows from thunderstorms may act as a source of pressure jump lines, or atmospheric internal bores.

3.1 THE MORNING GLORY

In the southern hemisphere, similar observations have been reported [4,5]. The most spectacular is the so-called "Morning Glory" which occurs in northern Australia near the southern coast of the Gulf of Carpentaria. This usually appears as a series of roll clouds, accompanied by a wind squall and a sharp rise in surface pressure. Fig. 3.1 shows its arrival from the east soon after sunrise. Several expeditions have been made to investigate the Morning Glory, and photographs have been taken which display the various forms of an internal bore which will be described in Chapter 13.

In the form "B" seen in the laboratory experiments in Chapter 13, rolls are very

Fig. 3.1 — Roll cloud marking the arrival of the Morning Glory at Burketown, northern
Australia, at 0630 local time on 12 October 1980. (Courtesy of Roger Smith).

clear, but turbulence appears in the rear face of the first roll and extends to those
which follow. A cloud formation displaying very similar behaviour is shown in Fig.
3.2.

As well as investigations by aeroplane, measurements of the flow patterns have
been made by pilot balloon ascents, and a typical example is shown in Fig. 3.3. This
shows the streamlines up to a height of 1500 m, deduced from seven balloon ascents
across a distance of 69 km. The thick line is the level of the top of the stable layer, and
shows that the strength (or height ratio) of the bore is about 2 and that the
wavelength of this undular bore is about 10 km.

The more intense Morning Glories have a strength much greater than that in this
example, and Fig. 3.4 shows one of these, passing Burketown in northern Australia
on 4 October 1979. It can be seen that (as in laboratory internal bores of strength 4 or
greater) the Glory now resembles a turbulent atmospheric gravity current; however,
the pressure pattern at the ground was still undular [4].

The origins of these disturbances are attributed to sea-breeze fronts interacting
with an existing nocturnal inversion. The Morning Glory which arrives from the
north-east is essentially formed by the collision of the east and west coast sea-breezes
[6,7]. Observations which can be interpreted as early stages in the formation of
internal bores by sea-breeze fronts will be dealt with in Chapter 4.

3.2 GENERATION OF BORES
The most probable generation mechanism for an atmospheric bore is the disturbance
of an existing stable layer by some kind of gravity current. The wide range
throughout which a gravity current can generate a bore advancing in a two-layer

Fig. 3.2 — The Morning Glory on 11 October 1981. Aerial view, showing a series of cloud rolls North-east of Burketown, looking east. (Courtesy of Roger Smith).

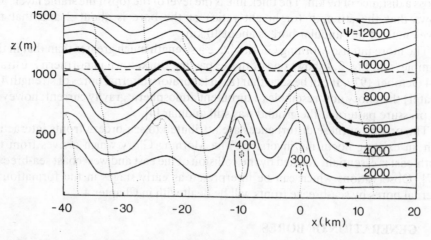

Fig. 3.3 — Morning Glory, 4 October 1979. Streamlines using seven pilot balloon ascents.

Fig. 3.4 — Morning Glory: turbulent form. (Courtesy of Derek Reid).

system will be examined in detail in the laboratory experiments of Chapter 13. Provided that the depth of the gravity current is greater than that of the stable layer, it can be shown that the bores formed will be similar for a step density change (as in the experiments) and for a linear density gradient which is more probable in the atmosphere.

Many of the gust fronts analysed by tower measurements have not fitted into the simple gravity current picture. In one series of 20 measurements [8] the fronts often consisted of multiple surges and it was necessary to distinguish between the onset of the cold air and the high momentum flow which defined the gust front, as these two did not always coincide.

3.2.1 Early stages of bore formation

The details of an early stage in the formation of a disturbance in a low level temperature inversion have been deduced from the pressure and temperature measurements from a 450 m tower [9] 6 miles north of Oklahoma City. The observations were made after the passage of an isolated thunderstorm. In Fig. 3.5(a) the ground level pressure is shown, and in Fig. 3.5(b) is plotted the potential temperature deduced from the tower data.

Immediately after the wind shift at 1625 CST there was a rapid pressure increase of about 1 mb, but no substantial change in ground temperature. This rapid pressure rise is seen to be associated with a wave propagating on the previously existing 100 m inversion. The temperature plot also shows a gravity current of cold air following the

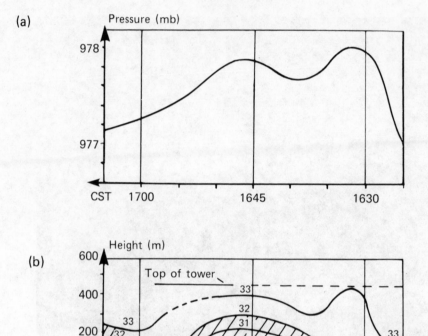

Fig. 3.5 — Observations made from an instrumented tower, showing an early stage in the formation of an atmospheric bore. (a) Ground level pressure. (b) Potential temperature, °C.

wave. Ground level pressure decreased as the wave passed over the tower but reached another peak at 1650 CST when the head of the gravity current was directly over the tower.

It appears from the temperature plot that a bore was in the first stages of generation when it passed the tower. It is estimated that the bore strength was about 3. The model described in Chapter 13 predicts a bore speed of $2(g'h)^{\frac{1}{2}}$ for this case, i.e. 8.1 m s^{-1}, which is close to the measured speed of the leading crest. The wavelength is deduced to have been about 7.3 km.

BIBLIOGRAPHY

[1] Christie, D. R., Muirhead, K. J. & Hales, A. L. 1979. Intrusive density flows in the lower troposphere: A source of atmospheric solitons. *J. Geophys. Res.* 84: 4959-4970.
[2] Tepper, M. 1950. A proposed mechanism of squall lines: the pressure jump line. *J. Meteor.* 7: 21–29.
[3] Schreffler, J. H. & Binkowski, F. S. 1981. Observations of pressure jump lines in the Midwest, 10–12 August 1976. *Mon. Wea. Rev.* 109: 1713-1725.

[4] Clarke, R. H., Smith, R. K. & Reid, D. G. The morning glory of the Gulf of Carpentaria: an atmospheric undular bore. *Mon. Wea. Rev.* 109: 1726–1750.

[5] Smith, R. K, Crook, N. & Roff, G. 1982. The morning glory: an extraordinary atmospheric undular bore. *Quart. J. Roy. Met. Soc.,* 108: 937–956.

[6] Clarke, R. H. 1984. Colliding sea-breezes and the creation of internal atmospheric bore waves: two-dimensional numerical studies. *Austr. Met. Mag.,* 31: 207–226.

[7] Noonan, J. A. & Smith, R. K. 1986. Sea breeze circulations over Cape York Peninsula and the generation of Gulf of Carpentaria cloudlike disturbances. *J. Atmos. Sci,* 43: 1679–1695.

[8] Goff, R. C. 1976. *Thunderstorm-outflow Kinematics and Dynamics.* NOAA Technical Memo. ERL NSSL-75. 69 pp.

[9] Marks, J. R. 1974. *Acoustic Radar Investigations of Boundary Layer Phenomena.* Rep. NAS8-28659. Dept. of Meteorology, University of Oklahoma.

4

Sea-breeze fronts

During the day in fine weather the sea-breeze blows from the sea towards the land; it is caused by diurnal temperature differences between land and sea. When the sun shines, the sea surface temperature changes very little, but the land becomes hotter and convection currents distribute heat throughout several thousand feet of air above the land. The sideways expansion of a column of air above the land, see Fig. 4.1, produces changes in pressure which are transmitted sideways with the speed of

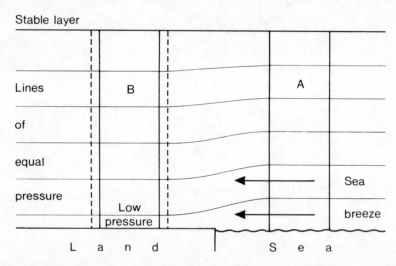

Fig. 4.1 — The development of the pressure field which gives rise to the sea-breeze at lower levels.

sound. The resulting pressure difference at low levels is responsible for the onset of the sea-breeze.

The sea-breeze caused by this pressure gradient at low levels is usually only 300–400 metres deep but the distance that the sea-breeze blows inland may extend considerably during the day. On a calm day the inland boundary of this spreading cool air is initially quite diffuse, extending over several kilometres.

On days when the sea-breeze meets an opposing wind, blowing towards the sea, a sharp boundary forms between the land-air and the cooler sea-air. A sea-breeze front then develops, which has all the characteristics of a gravity current of cold, dense air. A sea-breeze front can also form on calm days and even on days with a light onshore wind, but usually only much later during the day.

Several of the earliest aerial investigations of these sea-breeze fronts were made by meteorologists flying gliders, using the strong rising air at the front [1]. Sea-breeze fronts are also important to glider pilots in flight-planning because the fronts mark the boundaries of the much reduced convection in the sea-breeze itself.

The "sea-air" may extend over a large part of England on a fine summer day, and a good example of this is shown in Fig. 4.2. This shows the sea-breeze situation at

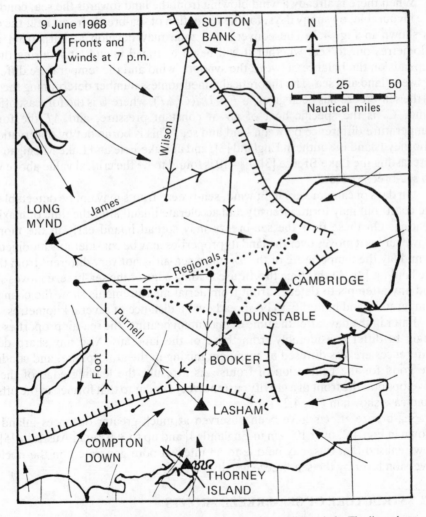

Fig. 4.2 — The spread of the sea-breeze in England on a day of light winds. The lines show tracks of glider flights in the area which remained convective.

1800 GMT on 9 June 1968, a good day for gliding and one of well-developed sea-breezes. Sea-breeze fronts had formed parallel to several coasts and by the time shown had travelled nearly 50 km. The map also depicts a number of memorable soaring flights made above the strongly convective countryside where the stable sea-breeze inversion layer had not yet encroached.

4.1 FORMATION OF SEA-BREEZE FRONTS

The day illustrated in Fig. 4.2 was one of very light winds, originally from the north-west. Any pre-existing wind system has a great influence on the development of sea-breeze fronts. For example, if the winds are already blowing from the sea towards the land then the strength of the diurnal sea-breeze effect is very much reduced.

When there is already a wind blowing from the land towards the sea, conditions are favourable, on sunny days, for the formation of sea-breeze fronts near the coast, as shown in Fig. 4.3. The convergence zone may develop a sharp front a few kilometres out to sea; whether it will begin to travel inland later during the day depends on the balance between the synoptic wind and the temperature difference between land and sea. The theoretical dimensionless number determining the onset of the sea-breeze has been given as $F_S = U^2/(C_p\Delta T)$, where U is the forecast offshore wind, C_p is the specific heat of air at constant pressure, and ΔT the forecast temperature difference between land and sea. This is borne out by observations at Thorney Island in southern England [2], and at Lake Erie and Lake Michigan when forecasting the Lake Breeze [3]. $F_S = 10$ is found to be the critical value above which no sea-breeze occurs.

On days of calm or very light wind, sea-breeze fronts may not be detectable near the coast, but may form, intensify and accelerate inland after the time of maximum heating. On these days the sea-breeze may spread inland early in the morning. However, as it moves over the land its properties may be considerably modified, and at midday the temperature of the foremost sea-air is not very different from that of the land-air [4]. At this time the diffuse leading edge of the sea-breeze moves slowly and any attempts to detect a front from aerial measurements show the changes of temperature and humidity to be spread over a distance of several kilometres.

Later in the day, after the time of maximum heating, the sea-air properties again begin to differ considerably from those of the land-air. Also any sharp density differences are less diffused by turbulent mixing in the atmosphere, and conditions are good for the generation of fronts. As a result, the leading edge of the flow develops into the front of a gravity current, very similar to that formed in the offshore wind case shown in Fig. 4.3.

Sea-breeze effects have been observed at much greater distances inland than shown in Fig. 4.2, over 100 km in England [4] and up to 300 km in Australia [5]. It is now realised that this may be due to an internal bore generated on the nocturnal inversion layer by the sea-breeze gravity current.

4.2 STRUCTURE OF SEA-BREEZE FRONTS

The advance of the sea-breeze inland in southern England has been monitored in a project based at the Lasham Gliding Centre. During a period of ten years, the sea-breeze passed Lasham, which is 45 km from the coast, on as many as 76 days [4].

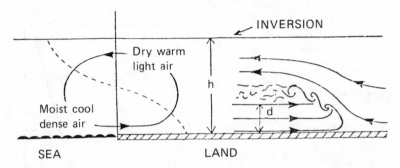

Fig. 4.3 — The development of a sea-breeze front, in a weak offshore wind.

Fig. 4.4 shows the progress of the sea-breeze inland on 14 June 1973, a summer day of light winds, and typical of many observations of deep inland penetration. The data points marked show the time of arrival of the sea-breeze as detected at ground stations measuring temperature, relative humidity and wind direction. Also marked are times and positions of pilot balloon ascents and two flights made in an instrumented light aeroplane. The times of passage shown here differ by less than an hour from the mean times of the 76 cases, although not all were traced so far inland after passing Lasham.

In the earlier stage, since there was no offshore wind to tighten up a front, the leading edge of the sea-breeze was diffuse and extended over 1 km or more. Later during the day, by 1600 GMT, a distinct front had developed, the position of which could be identified by very distinctive clouds forms, with base level much lower than that of the cumulus cloud further inland.

Observations of the structure of many sea-breeze fronts have shown that they behave like gravity currents of dense air. Typical measurements made from an instrumented light aircraft near such a front are shown in Fig. 4.5 [6]. The aircraft used was a Falke "motor-glider" which can be flown slowly and has many features similar to a glider. It carried a data-logging system which recorded height, airspeed, rate of climb, air temperature and water vapour density every 1.6 seconds, or every 50 m along the flight path. This plot of the humidity cross-section is built up from six horizontal traverses, made between 1820 and 1900 GMT, and shows the pattern near a sea-breeze front along a distance of 2 km, up to a height of 1 km.

The sea-breeze was 400 m deep, but the observations show moist air reaching about twice this height and the form suggests the formation of billows as seen in laboratory flows. Further similarities were shown by measurements of four other fronts under similar conditions, which showed an initial height of moist sea-air to be twice as great as that of the sea-air 10 km behind the front [6].

4.2.1 Cloud forms at sea-breeze front
The presence of a sea-breeze front can often be deduced from the ground by the formation of distinctive forms of cloud. On a clear day the rising air at the sea-breeze front may reach condensation height and form a line of cloud in an otherwise clear sky. On a day in which the sky is full of small fair-weather cumulus, the front may

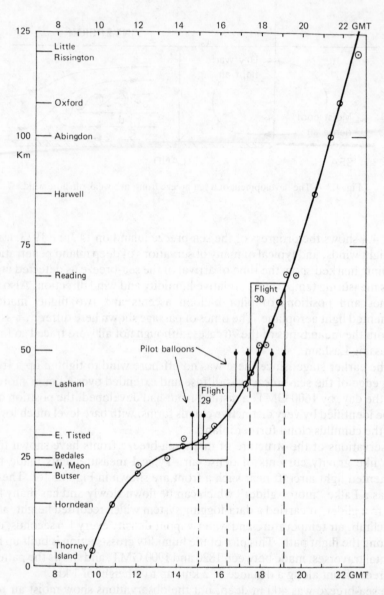

Fig. 4.4 — The progress of the sea-breeze inland on 14 June 1973, a sunny day of light winds. Before 1600 GMT the leading edge was diffuse; later a distinct front had developed.

make its presence visible both as a sea-ward limit to cumulus formation and by fragments of cloud with a much lower cloud-base. These clouds form as ragged veils or curtains and can be seen to be rising rapidly (in other respects they have the appearance of dissolving cumulus cloud). Fig. 4.6 is a view of a sea-breeze front seen from the air, flying just beneath the cloud-base of the cumulus clouds. Three short

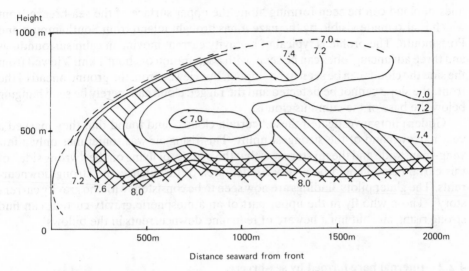

Fig. 4.5 — Humidity cross-section of a sea-breeze front, showing isopleths of mixing ratio in g kg^{-1}.

Fig. 4.6 — Clouds and haze at sea-breeze front near Selborne, 9 June 1968. (Courtesy of H. Howitt).

lines of cloud can be seen forming along the upper surface of the sea-breeze front which is also made visible by the haze it has brought inland from Southampton and Portsmouth. The profile is typical of a gravity current moving in calm surroundings and three advancing lobes can be seen, with a separation of about 1 km. Viewed from the side the clouds can be seen to be sloping, but viewed from the ground ahead of the front, the slope cannot be detected and the ragged cloud can merely be seen hanging below the base of the cumulus clouds.

Glider pilots investigated these "curtain clouds" and found that they marked a very narrow band of the strongest uplift. This rising air was sometimes only a few wing-spans wide and the mystery was that the penalty of flying on the "wrong side" of this curtain cloud could be a very sudden area of turbulence and strong downcurrents. The glider pilots' findings are now seen to be consistent with the gravity current story . Those who fly in the upper part of an atmospheric gravity current can find strong rising air, but must beware of fearsome downcurrents in the billows!

4.2.2   Internal bore formed by sea-breeze

One part of the world in which the progress of the sea-breeze has been extensively studied is South Australia, where sea breeze effects have been followed more than 300 km inland [6]. In the course of these studies it was found that the leading edge of the sea-breeze at sunset often appeared to be in the form of a "vortex" which sometimes separated from the following flow. Glider pilots flying at the sea-breeze front in southern England also noticed a marked change in the structure of the sea-breeze upcurrent near sunset. Instead of the rough, uncertain strip of lift found near curtain clouds earlier in the day, the rising air had spread to cover a much larger area and had also become very smooth. The characteristics had changed to those found by glider pilots when soaring in rising air displaced by atmospheric waves.

"Sea-breeze vortices" were subsequently found in England, and sometimes as many as three successive ones were traced, moving inland at 4 m s^{-1} with a spacing of about 10 km. The vortices can now be interpreted as being the early stages in the generation of an atmospheric bore by the movement of the sea-breeze front in the stable layer formed by the nocturnal inversion.

An early stage of the formation of an internal bore in the inversion has been measured north of Lasham and appears in Fig. 4.7. This shows the streamlines at the sea-breeze front just before sunset on 14 June 1973. It is clear that although the sea-breeze front in the early afternoon had been of gravity current form, the foremost part was by then almost separated from the following flow. The form resembles the first wave of an undular bore beginning to develop around the head of a gravity current as it moves through a stable layer, as illustrated in Chapter 13 in Fig. 13.7. The length of this section gives an estimate of 7 km for the wavelength of the undular bore.

This "sea-breeze bore" on 14 June 1973 was recorded at six more stations as it progressed inland. In the wind record at Harwell, 90 km from the coast, shown in Fig. 4.8, signs of waves are visible up to 3 hours after the first wave disturbance. The time for a wave to pass was about 45 minutes; at the measured velocity of 12 km h^{-1} this gives a wavelength of 9 km.

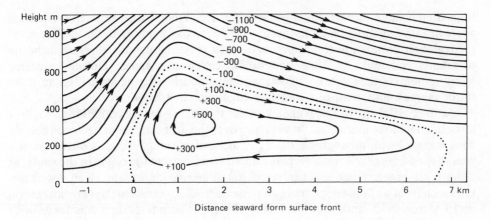

Fig. 4.7 — Streamlines at the sea-breeze front on 14 June 1973, just before sunset, showing the incipient bore.

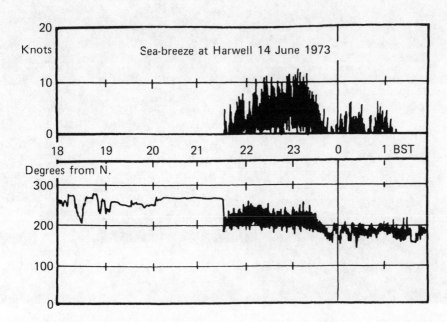

Fig. 4.8 — Wind strength and direction at Harwell on 14 June 1973. The oscillations are associated with the bore generated by the sea-breeze.

4.3 SEA-BREEZE AND POLLUTION

One of the most important practical applications of the study of sea-breeze fronts is the effect on the distribution of airborne pollution. At first thought it might seem that the effect of the sea-breeze is to enhance the dispersion of pollution, but actually it is

found that it may concentrate undesirable aerosols. Field work using tetroons, which
are constant pressure balloons used as tracers, has confirmed the existence of
recirculation patterns in the sea-breeze. Such recycling can concentrate airborne
pollutants and make possible undesirable chemical changes. During extensive
measurements of the sea-breeze flow in the neighbourhood of Chicago [7], one of the
tetroons was found to make a complete circulation in 3.5 hours.

At the coastal city of Los Angeles, under typical sea-breeze conditions a shallow
layer of cool moist marine air flows inland over the urban area and becomes heavily
contaminated with the ingredients of photochemical smog. About a dozen times a
year the sea-breeze reaches Riverside, about 60 miles from the origin of the smog, as
a sharp sea-breeze front between the polluted marine air and the clean desert air.
One such front was photographed as it passed across Riverside in the early afternoon
of 16 March 1972, and is shown in Fig. 4.9 [8]. The photograph was taken with

Fig. 4.9 — Front of sea-breeze polluted by photochemical smog, passing Riverside, California,
16 March 1972. (Courtesy of G. R. Stephens).

polaroid filter to darken the sky and increase its contrast with the cloud of polluted
air. The visible part of the smog cloud was about 1000 m deep and consisted of the
aerosol; it could not have been fog, since the relative humidity was only about 30%.

The characteristic feature of photochemical smog is its oxidising power, and part
of the oxidant record for this day is shown in Fig. 4.10. When the smog front passed

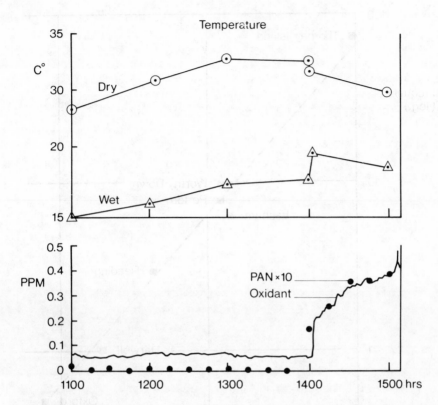

Fig. 4.10 — Records of wet and dry temperature and chemical concentrations at the front passing Riverside, 16 March 1972. PAN is peroxyacetyl nitrate. (After Stephens, 1975 [8]).

the air monitoring station the rise in oxidising power was so rapid as to be limited by the response time of the instrument. It can be seen that the value rose from 0.06 parts per million to a peak value of 0.43 ppm. Incidentally the US Federal air quality standard is 0.08 ppm.

A similar smog front used to be common in the Middlesbrough district on the north-east coast of England. During the morning a sharp smog front, very clearly marked by smoke haze, often crossed the town, reducing visibility to 100 m or less. Some ragged cumulus sometimes appeared at the front, with a condensation level much lower than that of the main cumulus clouds. However, with the reduction of pollutants in the air, this phenomenon is now much less common.

It appears that the air freshening that has been believed to accompany sea-breezes may be in some cases purely illusory, and pollutants may be concentrated continually by the sea-breeze, so that exceptionally high values can occur inland at sea-breeze fronts.

Some values of the frequency of inland penetration of sea-breezes have been measured in southern England, and the chart in Fig. 4.11 shows a summary made from continuous monitoring from 1962 to 1973. The logarithmic plot shows that the frequency falls off quickly with distance from the coast, reducing to less than one a year beyond 100 km inland.

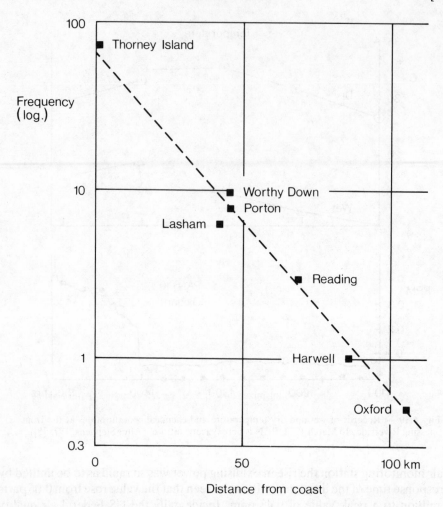

Fig. 4.11 — The frequency of penetration, in days per year, to points inland from the south
coast of England, averaged over a time of 12 years.

4.3.1 Sea-breeze remote sensing by lidar

The name "lidar" has been given to the technique of remote sensing in the
atmosphere by the use of light, usually a laser beam. Using the presence of aerosols
in the atmosphere as tracers, lidar has been used to measure the structure of a sea-
breeze front, as illustrated in the Frontispiece.

The lidar display shown in Fig. 4.12 is the vertical cross section of a sea-breeze
front, already illustrated in the Frontispiece, showing the concentration distribution
of aerosols [9]. This observation was made at 1740 hours, when the front, moving at
3.3 m s^{-1}, had reached a point 45 km inland from Tokyo, Japan.

The flow regimes shown in the figure consist of (1) the inflowing sea-breeze, of
depth about 300 m; (2) the mixing in the head, which reached a height of 1300 m; and
(3) the ambient flow. The details of the turbulent mixing behind the head are also
clearly shown.

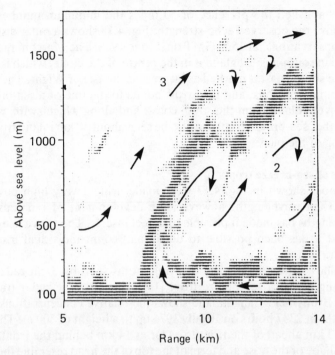

Fig. 4.12 — Analysis of vertical cross-section of a sea-breeze front seen by Lidar. (See false colour display shown in the Frontispiece.) The flow can be classified into three regimes: (1) the inflowing sea-breeze, (2) the returning mixed flow, and (3) the ambient wind.(After Nagane & Sasano, 1986 [9]).

4.4 BIRDS AND INSECTS AT SEA-BREEZE FRONTS

The bird which has become especially associated with sea-breeze fronts is the swift (*Apus apus*). Swifts are almost wholly aerial birds, picking up their livelihood, even their nesting materials, from airborne flotsam. It has been recorded [10] that on a fine day a single swift can catch up to 1000 airborne insects every hour. To continue at this rate all day when there are young to be fed, swifts are obliged to fly in places where there are a lot of insects. Glider pilots see swifts hunting for insects in thermals, and they have been noted in high concentrations flying along the line of strongest upcurrent at sea-breeze fronts [11].

Swifts catch a great variety of insects, the commonest being aphids and their allies. Aphids, which are among the most important agricultural pests in England, are only weak flyers but they exploit large-scale air currents for their distribution. As a sea-breeze front advances inland, swarms of aphids are lifted into the band of rising air ahead of the front, and swifts soaring at the front can find a continuous supply of their food.

4.4.1 Radar observations of sea-breeze fronts

The progress of an intense sea-breeze front moving inland in south-east England was first followed in 1961 [12]. On this occasion it was concluded that both moisture gradients and airborne swifts feeding at the front contributed to the radar returns.

Later work established the presence of 50 times the minimum number of swifts calculated to give the observed echo-strength. Fig. 4.13 shows a radar display made during these observations, at 1530 GMT on 2 June 1966. The Marconi radar station at Chelmsford in south-east England is in the centre of the circle, which is of 100 km radius, and the south coast is marked as far west as the Isle of Wight. The large dots are echoes from aeroplanes, and the arrow points to the line of a sea-breeze front which had moved inland from the south coast, and along which swifts were being counted. Another sea-breeze front can also be seen about 30 km inland from the east coast at the top of the picture.

4.4.2 Insects at sea-breeze fronts

The radar echoes shown in Fig. 4.13 were made using a very high power radar, working at 23 cm wavelength; however, the development of radar using much shorter wavelengths has made it possible to detect insects. They have been detected in clouds, and it has been possible to detect and count individual insects, from airborne radar.

Fig. 4.14 shows the area-density of flying moths seen by 3 cm radar from an aeroplane flying across a sea-breeze front [13]. This airborne radar traverse was made in New Brunswick, Canada, and shows the relative concentrations of spruce budworm moths at 2337 hours on 10 July 1976 up to a height of 500 m. The distance extends from 3 km ahead of the front to as far as 14 km behind the front. It can be seen that the shape of the nose and form of the top of the head resemble the form of a gravity current.

4.4.3 Insects at other small-scale fronts

Studies of the distribution of insect pests in Africa, particularly of the desert locust, have made clear the importance of weather systems in their life histories. For example, as a swarm of locusts approaches a coastline it has a good chance of meeting the sea-breeze which will prevent any possibility of being blown out to sea. The effect of sea-breeze fronts in concentrating the swarms also has an important influence on the behaviour of the insects.

Another important flying insect pest is the African armyworm moth, which flies at night. Radar detection of these insects has been used to trace the airflow in the atmosphere, particularly at fronts. During field work in Kenya [14], 3 cm radar was used to monitor individual moths flying within a distance of about 2 km. A cloud of moths, not so dense as to cause overlapping echoes, can reveal wind systems on a scale of 1 km or so by means of time-lapse photography of the radar screen, using one frame for each 3-second radar sweep. As the wind-shift line passed, marking the front of the cold outflow, it was possible to build up a vertical cross-section of echo motions. The outflow was clearly overtaking the wind shift as in the typical head of a gravity current.

There was no evidence that the strong concentration of moths was due to lifting of moths already flying densely near the ground, and it seemed probable that the moths were being concentrated by the convergent winds. The dramatic density increase that was observed needed only a descent of insects behind the front.

Locusts fly during the day, so the patterns formed by their swarms can be seen and photographed. Fig. 4.15 gives a striking view of a cold outflow outlined by flying

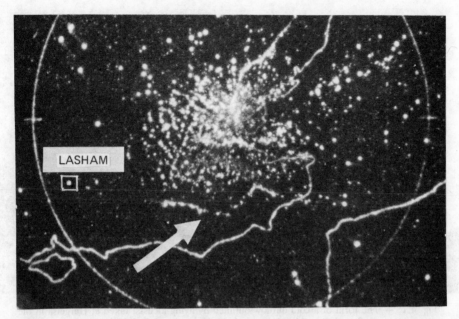

Fig. 4.13 — Sea-breeze fronts in south-east England outlined by radar, on 2 June 1966. The large dots are echoes from aeroplanes, and the smaller ones are from groups of birds. The arrow points to the author's aeroplane, flying along a line of swifts at a front. (Courtesy of Marconi Limited).

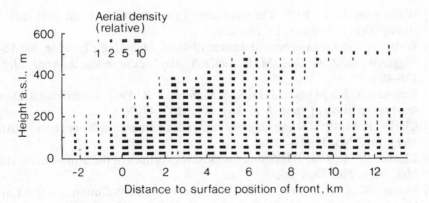

Fig. 4.14 — Density of flying moths at sea-breeze front, seen from airborne radar traverse, New Brunswick, Canada, 10 July 1976. (Courtesy of K. Allsopp).

Fig. 4.15 — The form of a cold outflow outlined by flying locusts, taken at Hargeisa in the Somali Republic, 3 August 1960. (Courtesy of A. J. Wood).

locusts. The pattern marked is similar to that shown in the radar measurements. The locusts can be seen moving in from the right and being swept upwards and backwards as they reach the front.

BIBLIOGRAPHY

[1] Wallington, C. E. 1959. The structure of the sea-breeze front as revealed by gliding flights. *Weather*, 14: 263–270.
[2] Watts, A. 1955. Sea-breeze at Thorney Island. *Met. Mag., London*, 84: 42–48.
[3] Biggs, W. G. & Graves, M. E. 1962. A lake breeze index. *J. Appl. Met.*, 1: 474–480.
[4] Simpson, J. E., Mansfield, D. S. & Milford, J. R. 1977. Inland penetration of sea-breeze fronts. *Quart. J. R. Met. Soc.*, 103: 47–76.
[5] Clarke, R. H. 1955. Some observations and comments on the sea breeze. *Austr. Met. Mag.*, No. 11, 47–68.
[6] Clarke, R. H., 1965. Horizontal mesoscale vortices in the atmosphere. *Austr. Met. Mag.*, No. 50, 1–25.
[7] Lyons, W. A. & Olsson, L. E. 1972. Mesoscale air pollution transport in the Chicago lake breeze. *J. Air Pollut. Control Assoc.*, 22, 876–881.
[8] Stephens, E. R. 1975. Chemistry and meteorology in an air pollution episode. *J. Air Poll. Control Assoc.*, 25: 521–524.
[9] Nakane, H. & Sasano, Y. 1986. Structure of a sea-breeze front revealed by a scanning Lidar observation. *J. Met. Soc. Japan Ser.II*, 64:(5) 787–792.
[10] Lack, D. 1956. *Swifts in a Tower*. Methuen, London, 239 pp.

[11] Simpson, J. E. 1967. Aerial and radar observations of some sea-breeze fronts. *Weather,* 22, 306–317.

[12] Eastwood, E. & Rider, G. C. 1961. A radar observation of a sea-breeze front. *Nature,* 189: 978–980.

[13] Schaefer, G. W. 1979. An airborne radar technique for the investigation and control of migrating insect pests. *Phil. Trans. R. Soc. Lond.,* B287: 459–465.

[14] Pedgley, D. E., Reynolds, D. R., Riley, J. R. & Tucker, M. R. 1982. Flying insects reveal small-scale wind systems. *Weather,* 37, 295–306.

5

Fronts and topography

The spread of a dense fluid on the rotating earth will be considered in Chapter 17, and will be illustrated by some laboratory experiments on a rotating table. Coriolis forces tend to oppose the spread of a dense fluid, but a very marked effect appears when a physical barrier is present. As the fluid turns to the right (in the Northern hemisphere) under Coriolis forces, nearly all the flow becomes concentrated into a band of dense fluid flowing parallel to the barrier. A vertical wall is not essential and a similar effect is seen with a sloping barrier.

This effect has been identified in the atmosphere in several parts of the world, and can be most marked when a large scale cold front approaches a coastline lying from north to south. If a range of mountains lies parallel to the coast, the flow of the dense air is slowed down and a strong, intense gravity current may develop, running parallel to the coast. Such trapped gravity currents form along the coasts of south-east Africa and south America and off the south-east coast of Australia.

It should be noted that in the Southern Hemisphere the Coriolis force causes a "turn to the left", hence the squally wind bursts from the south and moves along the east coast.

5.1 THE SOUTHERLY BUSTER

The southerly buster (or burster) is the local name given to a cold wind change which occurs in spring and summer along the coast of New South Wales in eastern Australia. The map in Fig. 5.1 shows the line of mountains which lies parallel to the coast and reaches a height of about 1500 m (5000 ft), and the area where the Southerly Buster appears. The squall line appears to be the front of a gravity current of cold dense air which is moving from the south and is held against the mountains by the Coriolis force.

The southerly buster is a mesoscale phenomenon, developing on a time scale of about a day, and the details of its behaviour are not observed by the regular synoptic network. The strength of the wind at a southerly buster is usually over 20 m s^{-1} and maximum gusts of over 37 m s^{-1} have been recorded [1]. Occasionally the leading

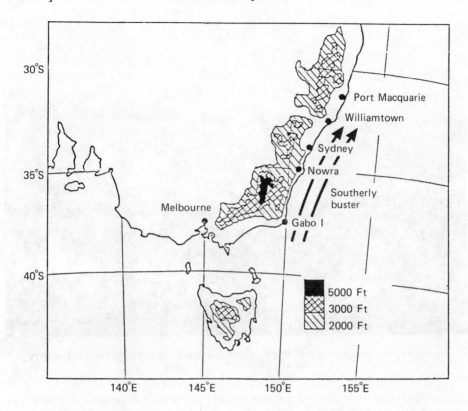

Fig. 5.1 — Map of south-east Australia, showing the path of the Southerly Buster.

edge of a southerly buster is marked by a spectacular roll cloud, aligned at right angles to the coast, as shown in Fig. 5.2. The front is not always accompanied by cloud and the abruptness of the wind change can be hazardous to low-flying aircraft and small boats. It does not usually rain at a buster, although rain often follows such a southerly change.

There can be dramatic temperature changes, with falls of 10–15°C in a few minutes in the afternoon. It has been found that the translation speeds along the coast are strongly affected by the local temperature and hence density difference across the front. It has been shown [2] that the speeds are well predicted by gravity current theory, incorporating this density difference and the depth of the cold air.

Wind records made during the spring and summmer show that when a front reaches the New South Wales coast it then advances northward much faster than its penetration inland [1]. The front does continue to move very slowly inland towards the higher ground, but its much faster steady progress along the coast in the form of a southerly buster is very clear.

The wind record made at Sydney Airport during the passage of a southerly buster is shown in Fig. 5.3(a). At 1900 local time the wind swung from the east to the south with a gust of maximum strength 23 m s^{-1} (about 50 mph) and blew at 10–15 m s^{-1} for about 15 minutes. It then increased and blew at over 15 m s^{-1} for 3 hours. This was one of the frequent (about 50%) occasions when pilot balloon ascents and

Fig. 5.2 — Roll cloud at Southerly Buster, seen from near the centre of Sydney, 16 Feburary 1985. (Courtesy of T. G. Donald).

surface wind records confirmed the presence of roll vortices in the post-frontal air, as indicated in Fig. 5.3(b). During several of the night-time passages, although the usual pressure increase of typically 2 mb was recorded, there was no appreciable temperature decrease, suggesting the arrival of an internal bore on the nocturnal inversion rather than a gravity current.

5.2 OTHER COASTAL TRAPPED GRAVITY CURRENTS

The low-pressure systems which move around the coast of southern Africa appear to have coastally trapped wave features which have something in common with some southerly busters [3].

A coastally trapped gravity current which moves from the south along the east coast of South America seems to have much in common with the southerly buster. Between Montevideo and Rio de Janeiro there is a range of mountains about 1800 m high, and a squall line called the Pompero Secco is known to move north, parallel to the coast [4]. Fig. 5.4 shows photographs taken from a ship, in which the top view shows the extremely smooth cloud form typical of the first wave of an undular atmospheric bore. The lower two show the presence of some "lobe" instability.

5.3 KATABATIC WINDS AND FRONTS

Mountains and valleys produce special winds which are much more complicated than those over flat country. Simple observations show that the local winds, close to the ground, change from upslope (anabatic) flow during the day to downslope (katabatic) flow at night.

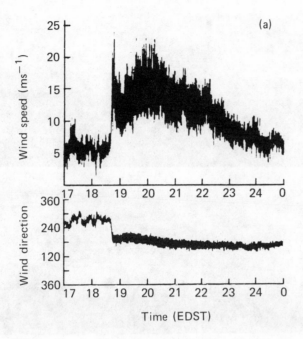

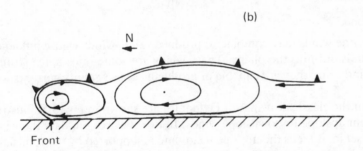

Fig. 5.3 — (a) Wind strength and direction at Sydney Airport on 11 December 1972. (b) Frontal structure in vertical cross-section parallel to the coast. (Courtesy of J. R. Colquhoun).

The diurnal mechanism which induces slope winds is similar to the mechanism we saw in Chapter 4 for the formation of the land/sea breeze circulation. A horizontal pressure gradient towards the slope is produced and an upslope wind results.

At night, the mechanism and the circulation are reversed, with the result that cold air drains down and away from the slope.

The upslope winds are not usually very strong, but their presence can be deduced from the formation of cumulus clouds on sunny mountain slopes. The currents which rise up slopes are used by glider pilots, but they need to fly very close to the mountain face to obtain this "lift". Measurements of slope upwinds have shown the maximum strength to be about 20 m from the ground, roughly equal to the wing-span of a glider.

Fig. 5.4 — Photograph of "pampero secco" off the east coast of South America. (From Georgi, 1935 [4]).

Downslope winds may combine to produce valley winds whose influence sometimes extends out into the plains. These winds are sometimes much stronger than upslope winds, and they are capable of producing dense gravity currents with sharp fronts.

Some of the classic drawings of Defant [5] are shown in Fig. 5.5, illustrating the change from upslope to downslope winds. In (a), towards midday, the wind up the valley has set in; it feeds the upslope winds and is supported by the return flow from the greater heights in the centre of the valley. In the late afternoon (b), the upslope winds have ceased to blow, but the valley wind continues for a short time. In (c) the downslope winds have started to blow, but since the valley wind is still flowing upslope, no cold flow down the valley results. In (d) the downslope winds alone remain, and in (e) the cold downslope air overcomes the up-valley flow and moves forward, sometimes as a sharp cold gravity current front. This onset of a katabatic wind as a sharp gust front is a common feature and at some places this onset has been described as an "air avalanche".

In some recent observations a different series of events was seen [6]. The temperature changes in the valley floor seem to have initiated the katabatic wind, which moved down the valley floor as a gravity current front. Here the katabatic wind was more likely to have been a result of the accumulation of cold air in the valley, which by reason of its greater density moved down the valley under gravity, followed by the downslope winds. Fig. 5.6 shows the situation at the mouth of Red Butte Canyon in the early evening.

The combination of several valley winds flowing from a mountain range may form

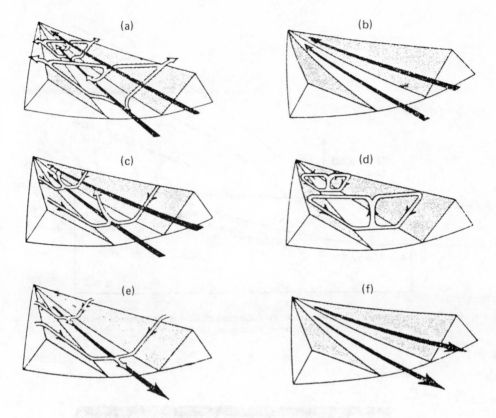

Fig. 5.5 — The development of down-valley winds. (After F. Defant).

a strong dense gravity current, resembling a cold outflow from a thunderstorm. In two cases of night-time cold air drainage down the eastern slope of the Rocky Mountains at Boulder, Colorado [7] the cold air progressed as a gravity current to the city of Boulder, 25 km distant. On both nights the drainage front occurred abruptly as a gravity current moving at about 3 m s^{-1}, the total depth of which was less than the 300 m height of the instrumented tower. Fig. 5.7 shows the acoustic sounder trace on one of these nights, showing the sharp front and a series of pulsations, resembling those seen in atmospheric bores, with a wavelength of about 10 km.

Very strong katabatic winds, known as "glacier winds", are formed from large ice sheets, and mostly continue throughout the 24 hours, with no interrupting upslope phase. The strongest of all are found around Antarctica, and in Fig. 5.8 is shown a wind record of a katabatic wind at Mawson, Antarctica, which after a strong gust-front reaches a strength of over 20 m s^{-1}.

BIBLIOGRAPHY

[1] Colquhoun, J. R. *et al.* 1985. The Southerly Burster of Eastern Australia. *Mon. Wea. Rev.* 113: 2090–2107.

[2] Coulman, C. E. *et al.* 1985. Orographically forced cold fronts — mean structure and motion. *Bound. Layer Meteor.*, 32: 57–83.

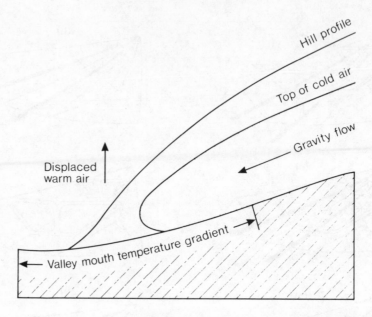

Fig. 5.6 — Initial stage of katabatic flow. Pressure gradient flow at the mouth of the Red Butte Canyon. (After Thompson, 1984 [6]).

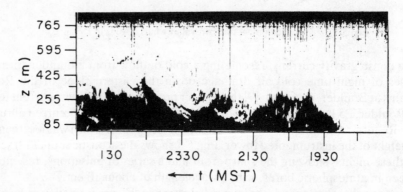

Fig 5.7 — Acoustic sounder trace, showing the arrival of drainage flow at Boulder, Colorado, on 8 October 1980. (Courtesy of W. Blumen).

[3] Gill, A. E. 1977. Coastally trapped waves in the atmosphere. *Quart. J. Roy Met. Soc.*, 103: 431–440.

[4] Georgi, J. 1935. Pampero secco von 17 Juli 1935. *Der Seewart*, 7: 199–205.

[5] Defant, F. 1951. Local winds. *Compendium of Meterology*, pp. 655–672. Amer. Met. Soc.

[6] Thompson, B. W. 1984. Small-scale katabatics and cold hollows. *Weather*, 41: 145–153.

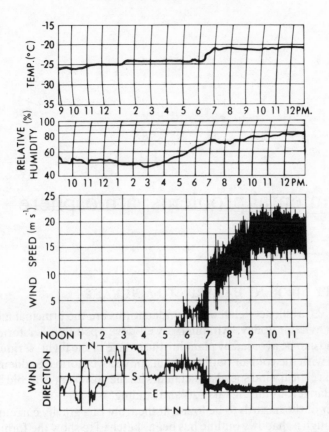

Fig. 5.8 — Wind speed and direction at onset of typical katabatic wind, 21 August 1960,
Mawson, Antarctica. (After Streten, 1963 [8], courtesy of Academic Press).

[7] Blumen, W. 1984. An observational study of instability and turbulence in
    nighttime drainage flows. *Bound. Layer Meteor.*, 28: 245–269.
[8] Streten, N. A. 1963. Some observations of antarctic katabatuc winds. *Austr.
    Met. Mag.*, 42: 1–23.

6

Environmental problems: atmosphere

6.1 GRAVITY CURRENT HAZARDS TO AIRCRAFT

Wind shear and turbulence related to thunderstorms are the principal meteorological hazards to aviation today. Although a well-developed thunderstorm is usually identifiable and can be avoided by planning the course of the flight, serious problems are associated with the cold air outflows from convective storms. Sudden changes in horizontal and vertical wind speed associated with the gust front of a cold outflow can be especially dangerous during take-off and landing.

A reminder of the hazards to be found at the front of a gravity current is given in Fig. 6.1. Although a shadowy outline has been sketched to show the form of the gust front, it must be realised that, unlike some other spectacular parts of the storm, the cold outflow is usually completely invisible. As noted previously, the leading edge of the gust front is maintained as a very sharp divide, so that the complete wind change may occur in a few tens of metres. As we follow this sharp interface to a height of a few hundred metres, it begins to roll up, forming billows. These grow until they become unstable and then decay, filling a layer several hundred metres deep with strong turbulence.

Aircraft have undergone structural failure when flying through the billow area, but the greatest hazard of flying through a single gust front is the almost instantaneous change in air speed which may be imposed. This change may be as much as 60 or 70 knots (30 or 35 m s^{-1}) in a few seconds, and there are times during both take-off and landing when the aircraft will not have this reserve above its stalling speed, with serious consequences.

A particularly difficult situation can occur when an aircraft, engaged in take-off or landing, flies right through a downburst cell in which the descending cold air is spreading out in all directions [1]. Fig. 6.2 shows events which have happened when an aircraft flies through both sides of a downburst during a landing or take-off. The aircraft making the approach in the upper diagram is flying at the correct airspeed and glide path when it meets a sudden headwind caused by the first part of the downburst; this results in an increase of airspeed and a climb above the correct glide-path. The pilot then reduces his airspeed and also steepens his approach. These

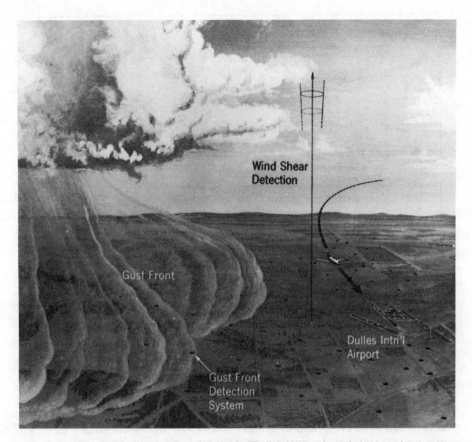

Fig. 6.1 — An invisible dense outflow approaching Dulles Airport through the detection system of pressure sensors. (Courtesy of American Meteorological Society).

procedures combine to make it impossible for him to cope with the tailwind and downcurrent which he encounters in the far side of the downburst. During a take-off through a downburst the hazards are very similar.

It is possible to reduce the dangers of flying through such cold outflows by warning the pilot of the presence of a downburst. Doppler radar has been used to give warning by detecting the approach of strong winds. A close network of pressure sensors has been set up around a few very busy airfields to detect the onset of dense flows from the pressure jump they cause [2]. There remains, however, the very serious problem of what have become known as "microbursts".

6.1.1 Microbursts

A microburst, as its name suggests, is a very small downburst of cold air from a thunderstorm. It is by no means small in the strength of the wind-shear and downcurrrent it can produce, but its shortness in duration, perhaps only 2–4 minutes, makes it very difficult to detect in advance. Microbursts have been studied by Professor Fujita of Chicago University [3]. He has recorded 18 aircraft accidents round the world which have been solely attributed to this phenomenon.

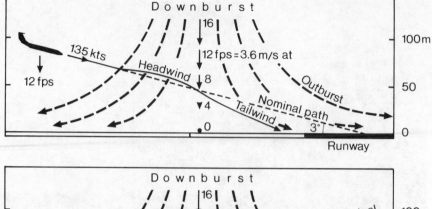

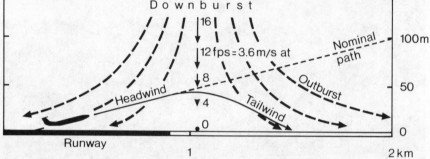

Fig. 6.2 — The effect of the airflow in a downburst on an aircraft approaching an airfield (above)
and taking-off from an airfield (below).
(after Fujita and Caracena, 1977 [7])

Measurements using doppler radar [4] and photographs of dust clouds and patterns in damaged crops and trees have led to a model as shown in Fig. 6.3. The observations show that an outflow microburst is surrounded by a ring vortex, spreading along the ground. This is exactly as seen in the very early stages of laboratory experiments in which a quantity of dense fluid is allowed to spread out along the ground. These experiments and the explanation of the formation of the vortex will be described in Chapter 12.

The descending air which forms a microburst is carried down by rain, and its density increases as it is cooled by the evaporation of water drops. If the cloud base is low the falling rain will not all disappear and the microburst will be "wet". In a "dry" microburst, all the rain has evaporated during the descent from a very high cloud base and an intense microburst is formed.

When the descending microburst air reaches the ground the outburst winds develop immediately after its touchdown. A strong vortex may be formed as shown in Fig. 6.4 where the form of the leading edge of a microburst outflow is made visible by dust raised in the strong wind.

6.2 SPREAD AND DILUTION OF A DENSE GAS

The number of accidents associated with accidental releases of dangerous heavy gases has grown during the past 10 years or so. For example, during this time the distribution and use of potentially dangerous liquid natural gas (LNG) has greatly increased.

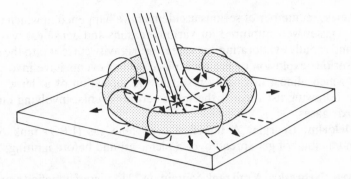

Fig. 6.3 — The ring vortex believed to form at the edge of an outflow microburst.

Fig. 6.4 — Photograph of the leading edge of a microburst showing the vortex outlined by a dust cloud. (Courtesy of Brian Waranauskas).

LNG was first developed in the USA in the 1930's and its use increased until an explosion in 1944 at Cleveland. Tanks containing 4200 m³ of LNG split and the cold vapour spread into streets, ignited and engulfed houses. One hundred and fifty people died.

After that, no new plants were built until the late 1950s when new safer tanks had been developed. Large-scale sea transport of LNG (up to 125 000 m³) and its storage in terminals adjacent to centres of population gave rise to concern about the consequences of an accident. Fortunately these fears have not been realised to date.

Nevertheless, a number of serious accidents have happened in which containers of liquid fuel gases were ruptured for various reasons and dense gas was released. The liquid and rapidly evaporating gas spread as a gravity current into the surroundings and a serious explosion or fire took place. Other events have involved highly toxic gases which almost invariably result in the formation of a dense, low-lying plume. The following list records the more serious events, involving either flammable or toxic gases.

1959; Meldrim, Georgia. A liquefied propylene gas (LPG) tank on a train ruptured and a cloud of gas spread over a picnic ground before igniting; 23 people died.

1969; Crete, Nebraska. A rail tank containing 72 tonnes of liquefied ammonia was completely ruptured in an impact following derailment; 9 people were killed.

1973; Potcherstroom, South Africa. An ammonia storage tank failed releasing 38 tonnes of gas; 18 people died.

1978; Los Alfraques, Spain. A road tanker carrying 43 m³ of LPG sprang a leak as it passed a camp site. The vapour spread over the camp and burst into flame; 150 people died.

1978; Highway near Mexico City. A road tanker spilt LPG after a collision; 20 people died.

1981; Montanas, Mexico. A train derailment resulted in fracture of tank cars containing chlorine. Over 100 tonnes of gas were released in a few minutes; 17 people died and 1000 needed hospital treatment.

1984; Mexico City. Explosion at a distribution centre for LPG adjacent to a shanty town. Over 500 people died.

1984; Bhopal, India. 40 tonnes of methyl isocyanate were released from a storage tank. The plume passed over a shanty town and about 2500 people died.

The need for the assessment of the hazards associated with such accidental releases of dangerous heavy (denser-than-air) gases has led to the development of a large number of simplified models of dense gas dispersion. The physics of some of the processes is not well understood and their description in the models involves a good deal of empirical input.

6.2.1 Dense gas dispersion models

Different types of mathematical models have been developed for the description of dense gases [5,6]; they can be separated into three categories: (A) fully three-dimensional, (B) depth -averaged, and (C) box models.

The three-dimensional models attempt to solve suitable approximations of the partial differential equations for conservation of mass, momentum and species to give the mean velocities and gas concentrations. The main differences between the models available [8], are not in the basic equations employed but in the numerical methods used to solve them. These models have the advantage that they can simulate a variety of release conditions, terrain and other obstacles present in the flow. They are expensive to run and uncertainties remain about the way the turbulence is modelled, justifying a simpler treatment.

Depth-averaged models further simplify the equations for the mean velocity and concentration by averaging vertically over the depth of the gas cloud. The turbulent mixing is then incorporated into the model as the rate at which the cloud height increases due to turbulent mixing. The main feature of these models is the treatment of the turbulent entrainment velocities. For example, in one model [9] it is assumed that the turbulent entrainment can occur at both the top and the edges of the cloud. One of the advantages over the fully three-dimensional group is the reduced cost of computation, but they maintain flexibility in computing cross-stream profiles.

"Box" models are far the most popular of all the types of models of dense gas dispersion [10,11], and general reviews have described in detail the different types [12]. Box models make the greatly simplifying assumption that the profiles of mean velocity and concentration in the cloud are known. The simplest box models assume that the cloud has the shape of a flat circular cylinder, with uniform height and radius in which the gas concentration is uniform. Some models assume different box profiles from the rectangular shape, but there is little advantage in this complication.

The schematic Fig. 6.5 shows a box model. The cloud is released at $t=0$, with its

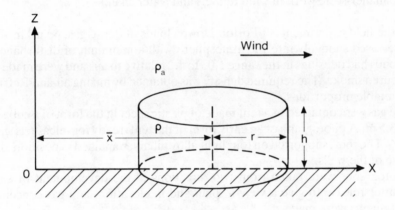

Fig. 6.5 — Sketch of a box model of a spreading cloud of dense gas.

centre at the point marked O, and is assumed to maintain a cylindrical shape, with height $h(t)$, radius $r(t)$ and uniform density (t) as it is advected downstream.

The equations governing the evolution of radius r, volume v, and x are:

$$v = \text{area} \times \text{height} = \pi r^2 h$$
$$dr/dt = \text{front speed} = c(g'h)^{\frac{1}{2}}$$
$$dv/dt = \pi r^2 u_z + 2\pi r h_r = \text{top entrainment} + \text{front entrainment}$$
$$dx/dt = u(z) = \text{wind speed at a height } z, \text{ whose specification varies from one}$$
$$\text{model to another.}$$

This model is only appropriate when the buoyancy forces are significant. Allowance has to be made as the stage is reached when atmospheric turbulence becomes the dominant mechanism for the dispersion of the cloud. In all the models, the specification of the entrainment velocities remains the chief question to be resolved.

To complement the theoretical modelling efforts, a number of laboratory and field experiments have been carried out. The relevant laboratory experiments will be described in detail in Chapters 12, 13, and 14. Two of the largest field trials performed in Britain have been at Porton Down and, more recently, at Thorney Island.

6.2.2 Porton Down and Thorney Island field experiments

The purpose of these large-scale experiments was to provide data which could be used for the following purposes:

(A)   to enable predictive models to be verified;
(B)   to further the understanding of the physical processes involved and to test model hypotheses;
(C)   to provide information on the scaling laws so that better uses could be made of smaller-scale work in wind tunnels and water channels.

In the field experiments at Porton Down clouds of dense gas, 40 m^3 in volume, were released suddenly into the atmosphere with a minimum of disturbance [13]. The clouds had densities in the range 1.03 to 4.2 relative to air, and were made visible by coloured smoke. The required density was obtained by mixing air and Refrigerant 12 in suitable proportions.

The gas was contained in a tent made of plastic sheet in the form of a cube of side about 3.5 m. A photograph of an early form of the tent, ready for release, is shown in Fig. 6.6. The roof, supported on four light alloy pillars, remained in position after the collapse of the walls.

Photographs were taken from two directions of the development of each gas cloud after release to give a view in plan and in elevation, and gas concentration measurements were made.

Each experiment at Thorney Island [14] was begun by releasing the much larger quantity of 2000 m^3 of a heavy gas nearly instantaneously from a cylindrical container into an ambient wind. The heavy gas was a mixture of nitrogen, Refrigerant-12 and smoke. In most of the experiments the mixture was adjusted so that it had a density about twice that of the surrounding air. The container was 14 m in diameter and 13 m high and was made of plastic sheeting whose sides could be drawn quickly down to the ground by elastic cords in about 2 seconds.

A second series of Thorney Island experiments was carried out in which the spreading dense gas met different types of fixed obstacles. These obstacles were a 5 m high solid fence, a 10 m high porous fence and a cube with sides 9 m high.

6.2.3 Release into calm air

Experiment 8 in the Porton series consisted of a release in the evening of a day when conditions were almost calm. The cloud collapsed symmetrically to the ground, and the vorticity generated at the cloud edge as it collapsed was concentrated at the front edge, forming a radially expanding vortex ring. It was apparent that at this stage most of the gas was contained within this "raised rim". Fig. 6.7 is a photograph of this effect, seen from above in one of the series of experiments at Thorney Island.

Fig. 6.6 — Early form of container for releasing a large volume of dense gas in a field experiment. (Courtesy of Health and Safety Executive).

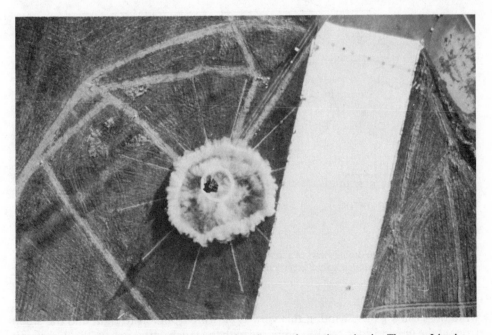

Fig. 6.7 — Release of dense gas into calm air seen from above in the Thorney Island experiments. (Courtesy of Health and Safety Executive).

6.2.4   Release into wind

After the initial collapse, the motion is mainly horizontal, as the front advances with the kind of gravity current head described in earlier chapters. The majority of the mixing with the ambient fluid occurs at the front of the current, and the density profile behind the head consists of a layer of roughly uniform density gradient, lying above a uniform density layer.

However, in an ambient wind the structures of the upwind and downwind fronts are quite different from each other. Fig. 6.8 shows a release at Thorney Island in a

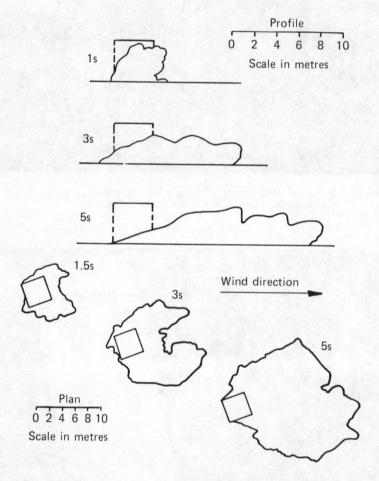

Fig. 6.8 — Plan and profile views of a dense cloud at various times after release in a 4.7 m s^{-1} wind in a field experiment. (Courtesy of D. G. Picknett).

wind of 4.7 m s^{-1}; the wind in this view is blowing from left to right and the steep front at the upwind edge can be compared with the much flatter one on the downwind edge. Comparisons made with the speeds of the upwind and downwind fronts at

Thorney Island are given in Fig. 6.9 in which the front speeds from the trials are compared with the empirical formula [15] to be derived later in Section 11.5. Stability considerations suggest that the upwind front is more stable than the downwind front, so more mixing would be expected at the downwind front.

During the gravity-spreading phase the stably stratified upper layer tends to suppress any turbulent mixing with the ambient fluid. However, in a turbulent environment there will be mixing due to the impinging external turbulence, and as the density driving effects diminish, the external turbulence takes over as the main control. As viewed experimentally this problem will be dealt with in Chapter 14.

6.2.5 Maplin Sands experiments

The Maplin Sands experiments were performed in the summer of 1980 [16]. As in the two series of experiments described above, the aim was to study the dispersion of dense gases, but there were important differences in the procedure.

A total of 34 spills of liquefied gas onto the sea surface was performed. The gases used were refrigerated LPG and LNG, in quantities up to about 20 m^3. An important feature of these experiments is the very large cooling of the air into which the liquid evaporates due to the latent heat required to cause the evaporation. In the later stages the main difference between the behaviour of propane and LNG spills is that propane remains denser than air, even if warmed up to ambient temperature. The behaviour of LNG is more complicated as the gas is initially dense due to its low temperature ($-162°C$) but heat transfer from the water and air will ultimately render the gas positively buoyant.

Release of the liquid was either continuous or instantaneous. In continuous spills liquid was released from the end of a tube near the water; for instantaneous spills the liquid was poured into a floating open-topped barge. When this was submerged the water flowed in, displacing the liquefied gas.

The site was a flat area of tidal sands on the north side of the Thames estuary and the releases were made 350 m from the shore at high tide during periods of offshore wind. An array of 360 instruments enabled measurements to be made of gas concentration, temperature and wind.

In one of the LNG spills, no.29, the supply was kept steady at 4.1 m^3/min^{-1} for 225 s into a steady wind of speed 7.4 m s^{-1}. The photograph of Fig. 6.10(a) shows the fully developed plume, extending over a length of 250 m.

The results of all the spills were compared with the predictions from the HEGADAS II model which was developed by Shell Research. Fig. 6.10(b) shows a plan view of the spill, together with the limits of the visible plume predicted by the Hegadas model, showing a close agreement.

6.3 GRAVITY CURRENTS OF GASES IN MINES

Extensive research has been done on the stably stratified layer of a contaminant gas which can form along the roof (or floor) in the ventilated roadways of coal mines [17]. In fact some of the earliest research on the behaviour of gravity currents, done in 1941 [18], was based on the free streaming of such a buoyant gas in a sloping channel.

Stably stratified layers have arisen in mine roadways when methane, which is lighter than air, is released from a source near the roof. If the stable density gradient

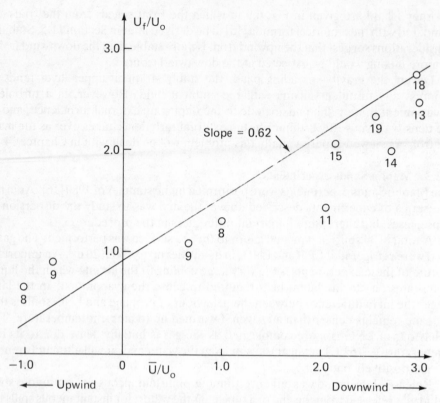

Fig. 6.9 — Plot of the upwind and downwind front speeds as functions of the mean wind speed for the Thorney Island releases. The solid line represents the empirical line developed by Simpson & Britter, 1980 [15].

of the gas becomes sufficiently large, mixing by turbulence will no longer be sufficient to disperse the gas. A layer can form and persist for long distances; if the ventilation is downhill a gravity current front will form. This front may move backwards, or become arrested, or it can even flow up the slope against the ventilation, as illustrated in Fig. 6.11. In cases such as this, knowledge would be required of the ventilation flow-rate which would ensure safe approach from the upwind side.

A large number of other environmental and industrial processes involve turbulent mixing in stratified flows; the important non-dimensional number concerned with this type of mixing is the Richardson number. This relates the restoring force on fluid when it is displaced from its equilibrium position to the forces available to mix the fluid.

To establish an overall Richardson Number, Ri_0, consider a layer of fluid depth H, moving between two horizontal planes, one above the other, of average density ρ, and total density difference $\Delta\rho$. Suppose the fluid at the lower plane moves horizontally with velocity ΔU relative to that at the upper boundary, and that the vertical gradients of density, $\Delta\rho/H$, and velocity, $\Delta U/H$, are constant. If we move a fluid parcel a small distance, d, upwards then the change in potential energy is

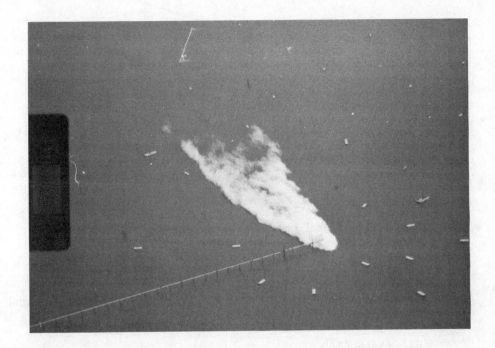

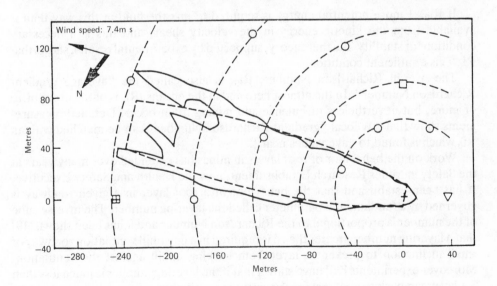

Fig. 6.10 — Release of dense gas in Maplin Sands experiment. The dotted line shows the limits predicted by HEGADAS II. (Courtesy of J. Puttock).

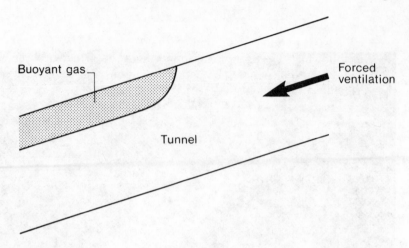

Fig. 6.11 — The effect of forced ventilation flow on the front of a buoyant gas on the roof of a tunnel in a mine.

$g\Delta\rho d^2/2H^2$, and the available kinetic energy is $\rho(\Delta U d)^2/2H^2$. The ratio of potential energy change to kinetic energy change is then

$$Ri_0 = g\Delta\rho H/\rho(\Delta U)^2$$

If $Ri_0 > 1$ more potential energy is required to mix the fluid in this way than is available from the kinetic energy in the velocity shear, and this is a necessary condition for stability. Further theory, supported by experimental results, shows that $Ri_0 > \frac{1}{4}$ is a sufficient condition.

The overall Richardson Number, Ri_0, is also called the "average gradient Richardson Number". In the atmosphere and in the oceans, Ri_0 is of the order of 10 or more, but nevertheless turbulence and mixing often occur. Detailed measurements show that any local "gradient Richardson Number" can be matched with an Ri_0 which is found to have values near $\frac{1}{4}$.

Work on the behaviour of roof layers in mines has extended over many years at the Safety in Mines Research Establishment, now the Health and Safety Executive. It has been established that the behaviour of a roof layer in a given roadway is governed by a fundamental parameter called the layering number. The inverse cube of the number is proportional to the Richardson Number and it has been shown [19] that a layering number greater than 3 is required in a downhill ventilation roadway of small inclination to prevent a layer from backing uphill against the ventilation. Moreover experiments [20] have shown that if the layering number is much less than 3 a layer can back a great distance against the ventilation.

6.3.1 Explosions caused by a layer or accumulation of methane
The following two examples of investigations give good illustrations of the behaviour of a buoyant gas in a complex situaton.

Cambrian Colliery

In 1965, 31 men were killed in an underground explosion at Cambrian Colliery. During the following investigations an opportunity occurred to observe another layer formed by reproducing the ventilation conditions that were believed to have existed at the time of the explosion [21].

Concentration measurements of methane were made at a series of points in the return ventilation roadway, which carried air from the mine-face. This roadway was 120 m long and sloped down at 1 in 33 from the face in the direction of the ventilation flow. It was believed that methane was entering the roadway from the roof between 90 and 110 m from the junction with the face.

The ventilation rate at the time of the accident was believed to have been either 0.15 or 0.1 m s^{-1}, and measurements were made with each of these flows, taking a time of half an hour for conditions to stabilise. The form of the methane observed near the roof under these two different conditions is shown in Fig. 6.12 (a) and (b).

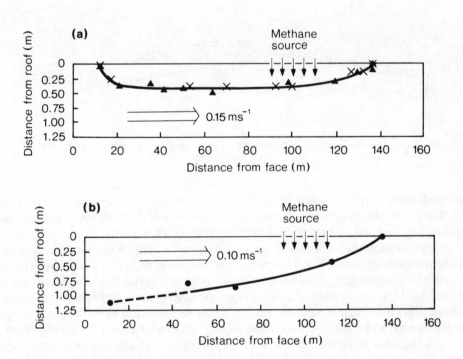

Fig. 6.12 — Measured methane concentration in Cambrian Colliery at two different ventilation speeds. (a) 0.15 m s^{-1} (b) 0.1 m s^{-1}. The curves show the 5% methane concentration. (Redrawn from Leach & Thompson, 1968 [21]; Crown copyright).

The contour shown is the 5% concentration of methane, and it can be seen that in the 0.15 m s^{-1} case the methane is contained by the ventilation flow, but in the case of the weaker ventilation much of the methane reaches the corner at the end of the mine-face. The methane distribution was also observed at a ventilation velocity of 0.38 m s^{-1}, when the methane layer dispersed in a few minutes.

Smoke was injected into the methane flow in both the low-speed cases and made it clear that the methane gas flow was backing against the ventilation. The layer number was found to be 0.7, much less than the critical number of 3. At such low layering numbers buoyancy forces predominate and a layer can back against the ventilation flow, and can become long in a short time.

It was calculated that there was a volume of about 24 m^3 of methane in the layer at the slower ventilation velocity, which is sufficient to explain the observed violence of the explosion.

Golborne Colliery

In 1979 an incident occurred at Golborne Colliery in which a large accumulation of methane was concerned [22]. An idealised diagram of the heading is shown in Fig. 6.13, showing the accumulation of methane in the upper part; this had occurred after

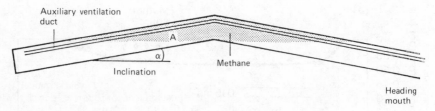

Fig. 6.13 — Idealised geometry of the heading in Golbourne Colliery. (From Mercer, 1981 [22]; Crown copyright).

a breakdown of the auxiliary ventilation. When an auxiliary fan was restarted the airflow caused the methane to be released at the mouth of the tunnel. Some of this gas was ignited and an explosion took place, killing ten workers.

One question that was raised was whether, on turning on the fan, the accumulation of methane would move *en masse* or would be eroded away more slowly by air channelling beneath it. A laboratory experiment was carried out, as illustrated in Fig. 6.14. This was performed, as are many described in the second part of this book, by an "inverted" experiment in which a dense salt solution took the place of the methane gas, and fresh water was pumped through two lengths of 50 mm glass tube.

All the tests were carried out with the interface between the brine and water initially as shown in Fig. 6.14(a). When water was introduced at the "face end" the brine accumulation moved down the tube as a plug. There was no tendency for the water to channel over the brine.

The plug motion continued until the leading edge of the interface reached the point where the tube changed from its downward slope to being upwards. The water then began to channel its way over the brine, forming a wedge of water above it, as shown in Fig. 6.14(b). The rate at which the wedge penetrated into the brine was about twice that measured in the earlier stages, suggesting that about half the cross-

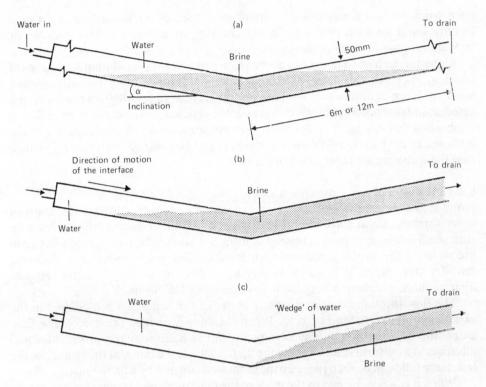

Fig. 6.14 — Laboratory model of the Colborne Colliery Incident. (a) Layout of model. (b) Motion of brine as a plug. (c) Water channelling over the brine once the dip has been cleared. (From Mercer, 1981 [22]; Crown copyright).

section of the tube was occupied by the water. Before water reached the outflow end of the tube, brine was pushed out at the same rate as the water entered the apparatus at the other end.

So, "re-inverting" the geometry, it was concluded that when methane had "locked-off" the heading, the accumulation would move as a plug once the fan was started. The rate at which the methane will leave the heading mouth would equal the rate of air input from the fan. This would apply, both before and after the air had broken through the crest, until the wedge of air reached the heading mouth. After that, a layer of methane would remain in the downward sloping leg of the heading, and would then be slowly eroded.

This type of work has been extended to include the study of buoyancy-driven exchange flows in damaged ducts, a flow which might arise in the coolant ducts of nuclear reactors. Some of the phenomena investigated in coal mines may also occur in vehicular tunnels and in underground garages.

6.4 HOUSE VENTILATION FLOWS

Very few houses in the world have artificially driven and controlled ventilation systems. Most rely on mixing by convection currents produced by heating from internal sources, and by open windows and doors for external ventilation. These open windows and doors can be responsible for excessive losses of heat in cold countries and invasion of hot air in hot climates, as gravity currents of dense or buoyant air pass through these spaces [23,24].

Flows through open doors can account for a significant part of total heat losses in housing. They can also determine the distribution of indoor contaminants within a building. Even at quite small temperature differences (a few degrees) between the exterior and interior air, buoyancy forces are significant and a gravity current flow is established through an open doorway. This flow may cause a loss of heat due to the intrusion of cold air along the floor, or a heat gain in a refrigerated room, in which case the incoming air flows along the ceiling.

6.4.1 A model of a two-dimensional flow

Flows set up in small-scale laboratory experiments serve to illustrate the kinds of flows which can occur and enable calculations to be made about the full-size flows. In such small-scale experiments, instead of using hot and cold air, it is usually better to use water as the working material with density differences produced by dissolved salt. By this means it is easier to maintain the correct range of the relevant dimensionless numbers concerned with viscosity and diffusion.

The flow through an open doorway at the end of a passage is modelled in the experiment illustrated in Fig. 6.15. When the door is opened the dyed dense fluid enters the building as a gravity current which fills half the depth of the door and advances at a uniform velocity along the floor. Outside the house at the same time the less dense fluid can be seen rising up the outer wall and mixing with the surroundings.

When it reaches the end of the passage the gravity current is reflected and can be seen travelling back towards the door. The dense fluid almost fills the space, but there is a small space above filled with lighter fluid which appears to have difficulty escaping due to frictional effects near the ceiling.

Experiments have also been carried out with an open staircase at the end of a passage, as shown in Fig. 6.16(a). The gravity current eventually fills up all the space in the ground floor; the dense fluid reaches the level of the upper floor but no higher. These conditions are shown in Fig. 6.16(a). Fig. 6.16(b) shows a similar physical process applied at the entrance to an Eskimo igloo. This consists of a tunnel through which a person can crawl; when he reaches the inside of the igloo he has to step over a barrier which is high enough to prevent the entrance of any gravity currents of cold dense air.

BIBLIOGRAPHY

[1] Fujita, T. T. 1981. Tornadoes and downbursts in the context of generalised planetary scales. *J. Atmos. Sci.* 38: 1512–1534.

[2] Bedard, A. J., Hooke, W. H. & Beran, D. W. 1977. The Dulles Airport pressure jump detector array for gust front detection. *Bull. Am. Met. Soc.,* 58: 920–926.

[3] Fujita, T. J. 1985. *The Downburst.* University of Chicago, 122 pp.

[4] Wakimoto, R. M. 1982. The life cycle of thunderstorm gust fronts as viewed with doppler radar and rawinsonde data. *Mon. Wea. Rev.* 110: 1060–1082

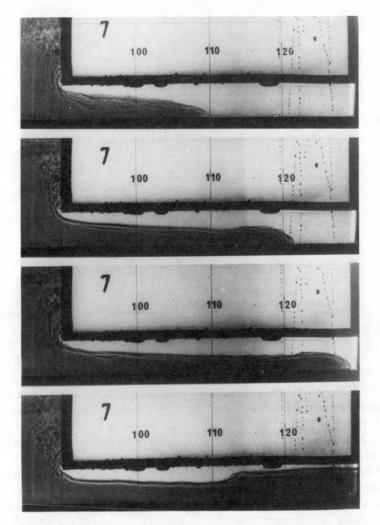

Fig. 6.15 — Laboratory model of dense flow entering a house. A gravity current of dense fluid flows through an open door into a passage between parallel walls.

[5] Blackmore, D. R., Herman, M. N. & Woodward, J. L. 1982. Heavy gas dispersion models. *J. Haz. Mat.* 6: 107–128.
[6] Havens, J. A. 1982. A review of mathematical models for prediction of heavy gas atmospheric dispersion. *Inst. Chem. E. Symp. on Assessment of Major Hazards*. University of Manchester.
[7] Fujita, T. T. & Caracena, F. 1977. An analysis of three weather-related aircraft accidents. *Bull. Amer. Meteor. Soc.*, 58: 1164–1181.
[8] Taft, J. R., Kyne, M. S. & Weston, D. A. 1982. MARIAM: A dispersion model for evaluation, realistic heavy gas spill scenarios, *American Gas Association Gas Dispersion Conference*, Chicago, Ill., 17–19 May 1982.
[9] Ermak, D. L., Chan, S. T., Morgan, D. L. & Morris, L. K., 1982. A comparison of dense gas dispersion model simulations with Burro series LNG test spill results. *J. Haz. Mat.* 6: 128–160

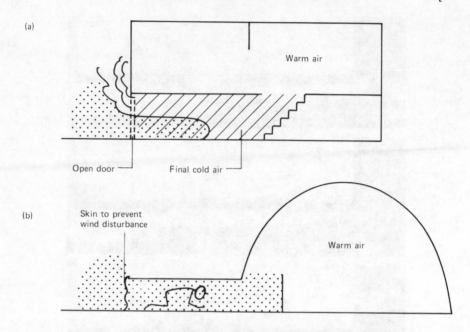

Fig. 6.16 — Comparison of (a) Western house and (b) Eskimo igloo.

[10] van Ulden, A. P. 1974. On the spreading of a heavy gas released near the ground. Proc. 1st Int.Symp. on Loss Prevention and Safety Promotion in the Process Industries, Holland.

[11] Fay, J. A. & Ranck, D. A. 1983. Comparison of experiments on dense gas cloud dispersion. *Atmos. Environ.*, 17: 239–248.

[12] Wheatley, C. J. & Webber, D. M. 1984. *Aspects of the Dispersion of Denser than Air Vapours Relevant to Gas Cloud Explosion.* Final Report of Contract between European Atomic Energy Community and the UK Atomic Energy Authority.

[13] Picknett, R. G. 1981. Dispersion of dense gas puffs in the atmosphere at ground level. *Atmos. Environ.* 15: 509–525.

[14] McQuaid, J. (ed.) 1985. *Heavy Gas Dispersion Trials at Thorney Island.* Elsevier Science Publishers, Amsterdam.

[15] Simpson, J. E. & Britter, R. E. 1980. A laboratory model of an atmospheric mesofront. *Quart. J. R. Met. Soc.*, 106: 485–500.

[16] Puttock, J. S., Colenbrander, G. W. & Blackmore D. R. 1984. Dispersion results from continuous releases of refrigerated liquid, propane and LNG. From *Air Pollution Modelling and its Applications III.* ed. C. D. Wispelaere. Plenum Publishing Corporation Ltd.

[17] Fletcher, B., McQuaid, J. & Mercer, A. 1973. Some ventilation problems in the underground environment. *Tunnels and Tunnelling,* 5: 144–150.

[18] Georgeson, E. M. H. 1942. The free streaming of gases in sloping channels. *Proc. Roy. Soc. London, Ser. A,* 180: 484–493.

[19] Bakke, P. & Leach, S. J. 1962. Principles of formation and dispersion of

methane roof layers and some remedial measures. *The Mining Engineer,* 22: 645–669.

[20] Bakke, P., Leach, S. J. & Slack, A. 1964. Some theoretical and experimental observations on the recirculation of mine ventilation. *Colliery Engineering,* 41: 471–477.

[21] Leach, S. J. & Thompson, H. 1968. Observations on a methane roof layer at Camborne Colliery. *Mining & Minerals Engineering* (August) 35–37.

[22] Mercer, A. 1981. An experimental investigation into the behaviour of a methane accumulation in an internal heading following auxiliary fan start–up. *19th Int. Conf. Research Institutes in Safety in Mines,* Katowice, October 1981.

[23] Brown, W. G. 1962. Natural convection through rectangular openings in partitions. *Int. J. Heat Mass Transfer,* 5: 859–868.

[24] Shaw. B. H. & Whyte, W. 1974. Air movement through doorways. *J. Inst. Heating & Ventilating Engineers,* 42: 210–218.

7

Gravity currents in rivers, lakes and the ocean

Sharp boundaries form in the ocean between adjacent water masses of different properties. These boundaries appear at the surface as frontal lines separating areas of water of different density due to differences in temperature, salinity or amount of suspended sediment.

Pioneering descriptive and dynamic studies of fronts were carried out [1] in the seas around Japan. Fronts were classified in terms of the different types of area where they might be expected to form, and much descriptive material was gathered from Japanese fishermen. For example, at the boundary between water masses they had often remarked upon a visible line of demarcation, with peculiar ripples. The sea could become quite violent in frontal zones, especially if the current shear across the front was large. Steep pyramidal standing waves could form when wave trains crossed into a region of impeding current, or where a strong current flowed in the opposite direction to a strong wind.

Flotsam had often been noticed to accumulate along the convergence line. This often included detritus such as dust, foam, timber and examples from the whole food chain up through plankton, mollusca, insects, fish, birds, whales and finally man (in the form of fishermen and also corpses!).

7.1 OCEAN SCALE FRONTS

Many oceanic fronts, with lifetimes of several months, are formed at the boundaries of two water masses having different origins, and on this large scale the flow is controlled by forces arising from the rotation of the earth.

Chapter 17 will discuss laboratory experiments on flows in a rotating tank, showing how the advance of the front of a gravity current is limited by Coriolis forces. The flow of buoyant fluid is eventually directed (in the Northern hemisphere) to the right, behind the front.

To give a large-scale example, the map in Fig. 7.1 shows a record of the position of the thermal front of the Gulf Stream during a 9-month period between February

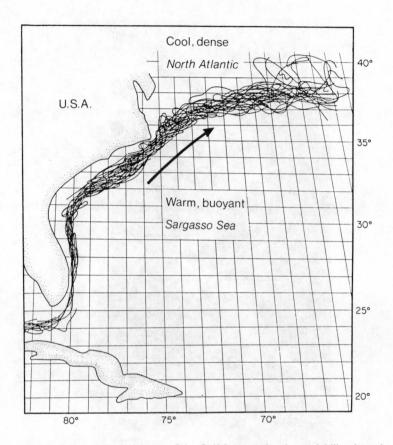

Fig. 7.1 — A composite of thermal fronts of the Gulf Stream, showing variability through a
9-month period. (After G.A.Maul).

and November 1978. These fronts are formed at the boundary of the spreading of the
gravity current of warm blue waters of the Sargasso Sea towards the North, above the
cold green waters of the North Atlantic. The Gulf Stream is the (less dense) warm
current near the surface flowing parallel to this boundary, partly due to rotational
forces and partly caused by the wind.

7.1.1 Shallow sea fronts
Small-scale counterparts of oceanic fronts are found in coastal waters and may result
from tidal mixing or from the interaction between fresh and salt water. Fig. 7.2 shows
a satellite photograph of south-west Britain in which differences in sea surface
temperature indicate two such fronts : one between Ireland and Cornwall and
another to the south.

The structure of a similar front investigated near the Channel Islands, both at and
beneath the surface [2], is shown schematically in Fig. 7.3. This shows two water
types A and B separated by a frontal region. Near the front there are convergent
motions which gather all floating material (w) to the edge of the front. Confirmation

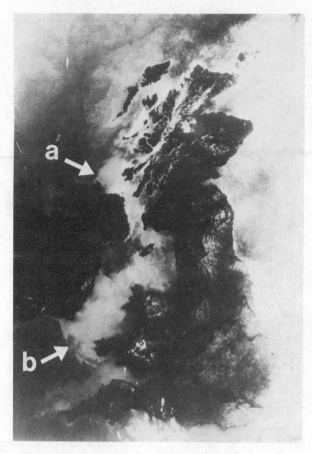

Fig. 7.2 — Satellite photograph showing fronts in the sea near Britain. (a) Islay-Malin Head
Front. (b) Celtic Sea Front. (Courtesy of Dundee University Electronics Laboratory).

of sinking motions was made by steaming along one side of the front and marking the
sea surface with a fluorescent dye. This line was seen to move towards the surface,
deform as shown and disappear below.

The flow in these observations appears to be that of a surface gravity current, as
described in Section 11.6. The less dense fluid, B, advances relative to the water mass
A forcing surface water beneath it. As the flow from both sides is towards the front,
any objects close to the surface will move towards this line, and if sufficiently buoyant
will collect there.

These shallow sea fronts or "shelfbreak" fronts are commonly formed at
boundaries between shallow nearshore waters, which are mixed by winds or tides,
and the deeper offshore waters which remain stratified. Fig. 7.4 is a simple diagram
showing the deep water which has become stratified due to solar heating down to the
thermocline, at a probable depth of about 20 metres. These shelf-sea fronts form in
sufficiently shallow water where tides generate enough turbulence at the sea-bottom
to prevent the formation of a seasonal thermocline.

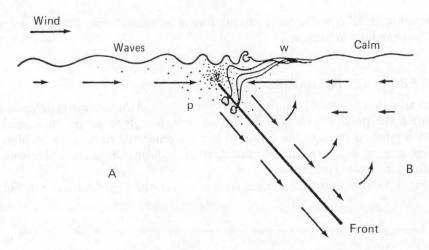

Fig. 7.3 — A schematic picture of the cross-section of a front separating two water masses A and B. (After Pingree, 1974 [2]).

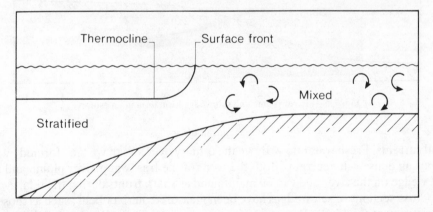

Fig. 7.4 — The formation of a "shelf-break front". The bottom-generated turbulence prevents the formation of the seasonal thermocline.

It has been shown that an important parameter whose value can be related to this kind of front formation is u^3/h, where u is the strength of the tidal stream and h the depth of the water [3]. The rate of work done by the turbulent kinetic energy is proportional to u, and the rate of change of protential energy it produces in mixing stratified water is proportional to the depth h. So in areas, as on the right in Fig. 7.4, mixed by strong bottom-generated turbulence, u^3/h is large. In the area on the left, u^3/h is small, and the area stratifies.

Critical values of this parameter have been shown to appear in the frontal areas which appear in the satellite photograph in Fig. 7.2, namely in the Celtic Sea and at the Islay-Malin front. Some of the problems of the dynamics of the formation and dissipation of this type of gravity current front in turbulent fields of various strengths will be examined in Section 14.12.

7.2 FRONTS IN ESTUARIES

In an estuary, the zone of transition between a river and the sea, many different flow regimes are possible. Large tidal stream velocities tend to produce estuaries vertically mixed through the action of bottom-generated turbulence. A high discharge rate can lead to significant stratification, reducing the amount of mixing and resulting in a two-layer flow.

Fig. 7.5 shows a possible configuration in a river with high discharge rate but low

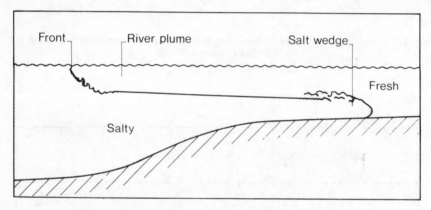

Fig. 7.5 — River plume and salt wedge formation in an estuary.

tidal currents. Fresh water flows down the estuary over a "salt wedge" formed by the opposing dense salty current. Both the front of the freshwater river plume and the salt wedge on the river bed can be maintained as sharp fronts during the tidal cycle.

Cross-sections of salt wedges have been measured on river beds, such as those in the estuary of the Fisher River, which opens into the Strait of Georgia [4]. Fig. 7.6 shows how, using an echo-sounding technique, the arrival of a salt wedge was recorded at a station near the coast. As the saline front moved past the station, it displayed a typical gravity current head.

The buoyant plume and dense wedge have been linked together in measurements made in the estuary of the River Seiont at Caernarfon in North Wales [5]. The freshwater discharge in this small river is forced back into the estuary during the flood phase of the tide, giving the appearance shown in Fig. 7.7. In this view the tidal inflow comes from the right and passes beneath the front of the river plume. Floating debris clearly marks out a characteristic V shape, in which two frontal arms meet at a point where rapid sinking motions occur. Surface flow relative to the front is towards it on both sides, and on reaching the front, surface fresh water sinks and is swept back upstream. When this photograph was taken the saline gravity current had already passed upstream along the river bed.

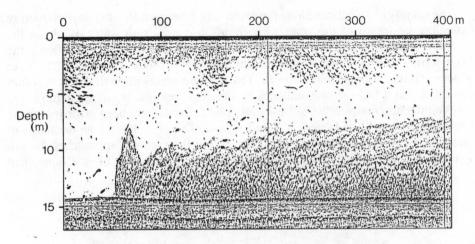

Fig. 7.6 — Salt wedge in the Fisher River, detected by echo sounder. (Courtesy of David Farmer).

Fig. 7.7 — Tidal intrusion front in the Seiont River, outlined by floating debris. Saline strait water is moving from right to left into the estuary. (Courtesy of R. Nunes).

The salinity distributions along the axis of the estuary show a two-layer structure extending upstream of the front with about 30 cm of fresh water overlying the intruding salt wedge. During the tidal cycle, after the time of maximum flood, the front starts to move downstream, maintaining the V configuration until the river widens and the front becomes convex. There are moments when the front becomes disrupted, but it has a strong tendency to return to the same organised pattern.

Well-documented studies of freshwater plume fronts have been made [6] for the Connecticut River, the largest in New England. This river has a high discharge of melt waters from mountains in the north, and in the spring it approaches the salt wedge case. Fig. 7.8 shows an aerial view of the frontal boundary of the Connecticut

Fig. 7.8 — Aerial view of the frontal boundary of the Connecticut River plume, 26 April 1972. River water is to the right.(Courtesy of R. W. Garvine).

River plume in April. The river is on the right of the foam line, which sharply divides the yellowish-brown river water from the blue-green waters of Long Island Sound. Measurements were made showing the vertical profile of density along cross-sections of the front, and an example is given in Fig. 7.9. This shows a typical gravity current profile, with the depth of the fresh water at this stage about half a metre.

7.3 FRONTS IN FJORDS AND LAKES

Fjords, or sea-lochs, are deep inlets from the ocean influenced by the tides, the winds and freshwater input from rivers. Many fjords have a very deep main basin, down to 1700 m, but most have a shallow sill at their mouth, between 1 and 100 m below sea level.

Several types of gravity current front have been observed in fjords [7]; Fig. 7.10 shows schematically some of the main processes which form them. Fjords with high river run-off have well-mixed layers which can be as deep as 10 m. When the tide is high enough, a dense gravity current may flow over the sill into this brackish water and run down the slope into the fjord. After it has descended some distance it may have the same density as its surroundings; it then starts to move horizontally, forming an interflow or intrusion as shown in the diagram.

The formation of a deep unmixed layer is of practical importance in connection with pollution. Deep-water exchange also determines the amount of available oxygen, which has important biological significance. The renewal of deep water due to the occasional flow of relatively dense water over the sill may occur only at rare intervals. For example, measurements in Loch Etive and in Loch Eil in Scotland have shown that the replacement frequency is only about once per year and that this is due to a gravity current process [8].

River fronts where extensive foam fronts are observed are permanent features of the surface waters of some fjords.

7.3.1 Lakes and reservoirs

A diagram of the flow of a river into a lake, or reservoir, is shown in Fig. 7.11. In this case the river is supposed to be carrying suspended material and is therefore denser than the water in the lake; it may also be colder and therefore denser still. At the entrance to the lake, the dense river water descends along a clearly marked line called the "plunge line". The diagram makes it clear that a plunge line is the sign of a stationary front on the surface, between the two water masses. As in surface fronts described above, this front may be detectable from colour changes in the water and by floating material which does not descend in the downward flow at the front. As the river inflow descends the slope in the form of a gravity current, mixing takes place at the head and eventually the fluid may move horizontally as an intrusion or inter-flow.

7.4 BORES IN THE ENVIRONMENT

Chapter 1 introduced bores and outlined some of the features that a bore has in common with the front of a gravity current. They both mark the leading edge of a continued process of mass transfer; this is not a typical feature of the behaviour of waves, the main effect of which is a transfer of energy.

Chapter 13 will examine the dynamics of bores and show how, at a comparatively weak bore, the loss of energy may be effected by a wave train, but that in a stronger bore there must be a turbulent mixing zone at the front. The latter process can be

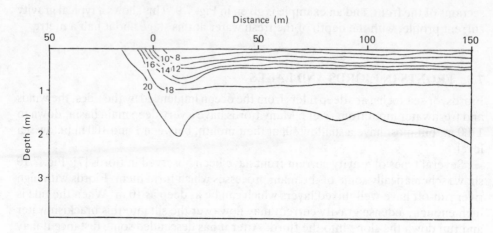

Fig. 7.9 — Density profiles in parts per 1000. Normal to the colour front in the Connecticut River, 14 May 1973.

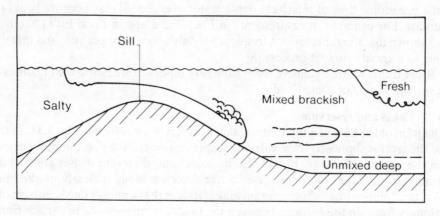

Fig. 7.10 — Some of the main processes which form gravity current fronts in a fjord.

seen at the leading edge of a gravity current (except in very viscous flows), the leading edge being almost identical to that of an internal bore at an interface between two fluids.

7.4.1   Tidal bores in rivers
We showed a picture in Chapter 1 of the bore moving up the River Severn. This is the best known bore in Britain, and is usually well developed at the spring tides in the spring and autumn equinox, when it can be as much as 1 metre high and may move up the river past Gloucester at speeds between 5 and 7 m s^{-1} [9].

Whether a bore occurs in a river depends on the local conditions at the mouth; two of the most important requirements are a very high tidal range and a funnel-

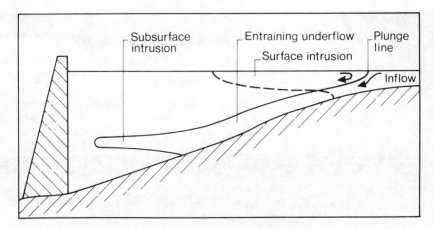

Fig. 7.11 — The flow of dense river water into a reservoir or lake.

shaped estuary. Other important factors are the rate of flow of the river and the shape of the river bed [10,11].

In Britain there are several other rivers which sometimes produce bores. After the Severn, the Trent usually produces the next most impressive bores. Here the tidal hydraulic jump is named the "aegre" and the waves following it are called "whelps" [12]. Clearly detectable tidal bores have been seen in several other rivers, although they may sometimes be only a few inches high.

Over 50 rivers throughout the world are known to have well-formed tidal bores. The best developed is probably the one in the Araguari River, on the coast of Brazil, which has been investigated by a team of scientists aboard the Cousteau Society vessel Calypso [13]. The photograph in Fig. 7.12 shows this to be an undular bore, with over 20 waves clealy visible behind the front of the bore.

Some of the South American bores have only recently been explored, but in China one of the largest bores in the world has historical records going back over 2000 years. The Qiantang River in China, which enters the sea only 50 miles south of the Yangtse, has a bore with a legendary history going back to the 5th century BC. Already in the 2nd century BC the Chinese understood its nature and expected a good bore at full moon.

As the city of Hangzhou grew up between the West Lake and the Qiantang river, its safety became more and more precarious and in 910 AD Governor Qian Liu built a dyke to meet the tide-water. He made bamboo bands, piled huge stones and drove in large trees. In addition he stationed hundreds of crossbowmen to shoot arrows, nominally "to stay the forces of the tide", but perhaps more effectively as a public relations activity. An old print reproduced in Fig. 7.13 gives an impression of the scene as they began to build the wall, parts of which still stand today. The bore can be seen advancing as a turbulent line from the left.

Throughout the following 1000 years the bore continued to trouble the people of Hangzhou. It seems to have been somewhat variable in strength, since Marco Polo did not remark on it during his visit.

In the 19th century careful records were made by a British Naval officer [14] who

Fig. 7.12 — Tidal bore at the mouth of the Araguari River, in Brazil. The bore is undular, with over 20 waves visible. (Courtesy of D. K. Lynch).

Fig. 7.13 — Protection against the bore in the Qiantang River: a historic scene.

recorded bores up to 2 metres in height. He noted how navigation was effected by the use of "bore shelters". These consisted of shelves half way up the river wall, well above the water level, on which vessels were perched until the bore arrived and floated them off.

Less than 20 years ago, although work had been done on maintaining the walls and digging out the channels, severe bores still sometimes occurred in the Qiantang river. An example is shown of one of these in Fig. 7.14, (a) and (b). In these views a

(a)

(b)

Fig. 7.14 — Intensification of a bore by reflection. In (a) the Qiantang bore has been reflected from the wall seen to the left. In (b) the original and reflected flow meet at the bend in the wall. (Courtesy of Zhejiang Provincial Estuarine Research Institute).

special feature is the dangerous intensification which can occur as the result of reflection of a bore which strikes the wall at an angle. The turbulent bore approaches from the top right-hand side and is reflected from the wall on the left. After reflection the height of the bore is doubled and it strikes the obstacles in the foreground with great force. Work carried out by the Zhejiang Provincial Estuarine Research Institute has reduced the effects of the bore; this has been done by deepening the river channels.

The highest tides in the world occur in the Bay of Fundy in Canada and, as might be expected, some well-known river bores are associated with them [15]. One of these rivers, the Petitcodiac, has a very impressive bore which may be seen to best advantage from the riverside park provided for this purpose in the city of Moncton, New Brunswick.

7.5 INTERNAL BORES

There is good evidence for the existence of internal bores in rivers, lakes, fjords and the oceans. Echo sounders can display the bore profiles and detailed measurements have been made using sensors of temperature, salinity and velocity.

These bores are usually generated in a comparatively thin layer of water which lies above a deeper denser layer; so the front of the bore would be expected to appear as a wave of depression. Several different mechanisms have been established for the formation of internal bores in these commonly existing stratified layers.

7.5.1 Surges in stratified lakes

Internal bores have been observed in Loch Ness, which is 35 km long, about 2 km wide and 200 m deep [16]. The bores propagate in undular form at speeds up to 0.2 m s^{-1} towards the north-west. They are often 10 m deep, with a series of waves to the same depth, about 1.0 km long.

The form of the bore is best shown in temperature measurements as in Fig. 7.15.

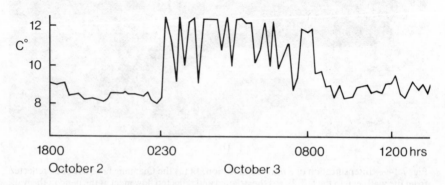

Fig. 7.15 — Temperature at 40 m depth in Loch Ness during passage of internal bore. (After Thorpe *et al.*, 1972 [16]).

This record was made from a moored boat from which a probe was suspended at a depth of 40 m near the centre of the loch. The increase in temperature as the front of the bore passed shows that the isotherms were lowered and that the bore was a wave of depression. The change in level of the isotherms was about 12 m and the mean speed of propagation of the bore was found to be 0.37 m s^{-1}. It is believed that this bore is forced by wind stress causing flow of near surface water to one end of the loch.

Even larger undular bores have been measured in Seneca Lake, New York [17], where the isotherms are often 20 m deeper after the front of the bore which travels at 0.35 to 0.40 m s^{-1}.

7.5.2 Internal tidal bores in the ocean
Internal bores have been investigated off the coast of California [18]. During the summer months, when a strong seasonal thermocline exists, the thermal structure approximates to a two-layer system. It was found that as an internal tide moved inland its wave profile became increasingly asymmetric as the wave entered shoaling coastal waters. This asymmetry became more pronounced with increasing wave height and resulted in the formation of an internal tidal bore.

In other measurements made at La Jolla, California [19], similar surges of cold water were recorded. Since the surge was evident even on the bottom it was believed to be due to the run-up of broken internal waves.

7.5.3 Flows over sills or through contractions
The presence of a sill near the mouth of a river or fjord may be responsible for the formation of an internal bore at certain times during the tidal cycle. In many cases a response to tidal flow over a submarine sill is an internal hydraulic jump downstream of the sill [20]. Such a flow occurs in Knight Inlet, British Columbia, on the turn to flood tide when an undular bore emerges from the region of the sill. Fig. 7.16 shows a

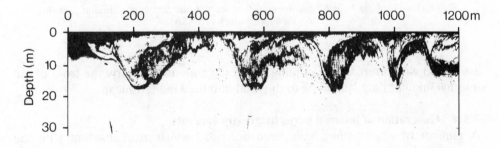

Fig. 7.16 — Echo-sounder record of an undular internal bore in Knight Inlet. (Courtesy of David Farmer).

record from a 200 kHz hull-mounted echo-sounder, taken as the ship traversed the bore. The leading edge of the undular bore is to the right of the picture and typical wavelengths are of the order of 100–200 m near the leading edge. At this internal jump the height ratio, or strength, is about 2.

Undular bores form at certain tidal stages in the two-layer flow through the Straits of Gibraltar [21]. Since the effect of evaporation exceeds that of river discharge in the Mediterranean, the water is more saline than that of the Atlantic. As the tide rises, a layer of less saline water flowing above a sill in the Strait enters the Mediterranean as an internal bore. The satellite photograph, Fig. 7.17, shows interfacial features

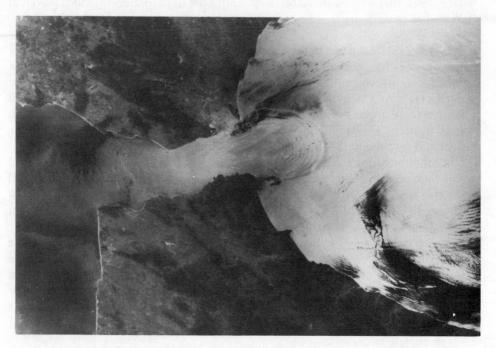

Fig. 7.17 — Undular bore formed by tidal flow through the Straits of Gibraltar. Satellite photography, (Courtesy of NASA).

associated with internal flow glinting in the sun and marks clearly the front of the internal undular bore spreading to the right into the Mediterranean.

7.5.4 Generation of internal bores by gravity currents

A number of observations have been described which are consistent with the generation by gravity currents of internal bores near water surfaces. Unfortunately many of these are not well documented.

Developing sets of lines at the surface suggesting an undular bore have been recorded on several occasions. Some of these have appeared as foam lines on the surface and others, from airborne thermal imagery, have shown periodic lines of variation of surface temperature.

Figs 7.7 and 7.8 have shown how the convergence zone at the front of a gravity current of less-dense fluid can cause a single sharp line of floating foam and debris. Laboratory experiments can reproduce the early stages in the formation of a bore at the front of a gravity current. As the front of the bore moves along the surface of the water ahead of the leading edge of the gravity current it will develop its own separate foam line. This process will be repeated until a series of parallel foam lines is formed. Fig. 7.18 is a photograph of a regular series of foam lines in Trondheim Fjord [7], which is thought to have this explanation.

Fig. 7.18 — A regular series of foam lines seen in Trondheim Fjord. (Courtesy of Norwegian Hydrotechnical Laboratory).

A similar pattern, suggesting an internal bore, has also been found in freshwater river discharges into the sea. Cold fresh water from the Quinault River discharging into the north-east Pacific shows a series of lines which have been shown to be successive boundaries of warm coastal-ocean water [22].

7.6 TURBIDITY CURRENTS ON THE OCEAN BED

The deep ocean floor is sometimes violently disturbed by turbidity currents moving down from the continental shelf. The attention of oceanographers was first drawn to this fact by breaks in submarine telephone cables, but the suggestion that turbidity currents were responsible for these breaks and other phenomena such as submarine canyons formerly created much controversy.

Some enormous submarine "landslides" are known to have set up turbidity currents at the edge of the continental shelf. The volume of sediment deposited by the well studied turbidity current off the Grand Banks near Newfoundland in 1929 has been estimated to be as much as 100 cubic kilometres. The events started with a severe earthquake which immediately broke a large number of submarine cables near its epicentre on the continental shelf just south of Newfoundland. But none of the numerous cables crossing the continental shelf in the area further south were affected. Further downstream from the epicentre, however, another cable broke an hour later. Then four other cables broke in succession, each in two or three places, the last break occurring 13 hours later and 300 miles distant. The details are shown in Fig. 7.19.

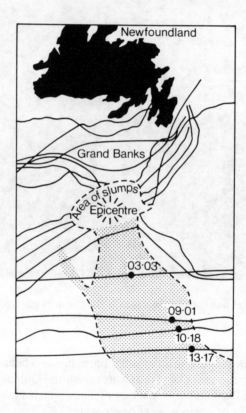

Fig. 7.19 — Submarine cables broken by a turbidity current from the Grand Banks sediment slide, 1929.

Submarine canyons are found in many sections of the continental slope. They are similar to river gorges on earth, with equally steep sides and depths. Turbidity currents have been observed actually flowing down some of these canyons. A good example is the event in 1935 at the mouth of the Rio Magdalena in Colombia: 450 m of a breakwater suddenly disappeared into the sea , and the same night a submarine cable was broken at a point 24 km out to sea and 1.5 km deep in one of the canyons which extend out to sea from near the river mouth. During repairs the cable was found to have grass wrapped round it of the type growing near the breakwater, suggesting that the slump had developed into a turbidity current capable of producing this long-distance transport. In this district during the first 25 years since one of the cables was laid it has been broken by turbidity currents on 17 occasions.

Abyssal plains occur in many oceans, for example they cover 10% of the floor of the North Atlantic, deviating from smoothness only by a few metres. Echo soundings show, however, that the smooth plains are themselves underlain by a rugged base. Analysis of cores from these plains shows that they consist of sand and silt layers which appear to have come from the continental shelf, and that transport by turbidity currents is the most likely explanation.

BIBLIOGRAPHY

[1] Uda, M. 1938. Researches on "Siome" or current rip in the seas and oceans. *Geophys. Mag.*, 11: 307–372.
[2] Pingree, R. D. 1974. Turbulent convergent tidal fronts. *J. Mar. Biol. Assoc. UK*, 54: 469–479.
[3] Simpson, J. H. & Bowers, D. 1981. Modes of stratification and frontal movement in shelf seas. *Deep-Sea Research*, 28A: 727–738.
[4] Geyer, R. 1983. *Fraser River salt wedge investigation*. Prelim. Report, University of Washington, Seattle, 13 July 1983.
[5] Simpson, J. H. & Nunes, R. A. 1981. The tidal instrusion front: An estuarine convergence zone. *Estuarine, Coastal and Shelf Science*, 13: 257–266.
[6] Garvine, R. W. & Monk, J. D. 1974. Frontal structure of a river plume. *J. Geophys. Res.*, 79: 2251–2259.
[7] McClimans, T. A. 1978. Fronts in fjords. *Geophys. Astrophys. Fluid Dynamics*, 11: 23–34.
[8] Edwards, A., Edelsten, D. J., Saunder, M. A., & Stanley. S. 0. 1980. Renewal and entrainment in Loch Eil, a periodically ventilated Scottish fjord. In *Fjord Oceanography*, ed. H. J. Freeland *et al.*, NATO Conference Series IV, Marine Science, Vol. 4. pp. 523–534.Plenum Press, New York.
[9] Rowbotham, F. W. 1964. *The Severn Bore*. David & Charles, London, 100 pp.
[10] Tricker, R. A. R. 1964. *Bores, Breakers and Waves*. Miller & Bonn Ltd, London.
[11] Lynch, D. K. & Bartsch-Winkler, S. 1986. *Tidal Bores*. Unpublished report.
[12] Barnes, F. A. 1952. The Trent Aegre. *Survey, (University of Nottingham)*, 3: 1–16.
[13] Murphy, R. 1983. Pororoca! *Calypso log*, June 1983, 8–11.
[14] Moore, R. N. 1883. *Report on the Bore of the Tsien-Tang-Kiang*. Potter, London.
[15] Dalton, F. K. 1951. Fundy's prodigious tides and Petitcodiac's tidal bore. *J. Royal Astron. Soc. Canada*, 45: 225–230.
[16] Thorpe, S. A., Hall, A., Crofts, I. 1972. The tidal surge in Loch Ness. *Nature*, 237: 96–98.
[17] Hunkins, K. & Fliegel, M. 1973. Internal undular surges in Seneca Lake. *J. Geophys. Res.*, 78: 539–548.

[18] Cairns, J. L. 1967. Asymmetry of internal tidal waves in shallow coastal waters. *J. Geophys. Res.* 72: 3563–3565.

[19] Winant, C. D. 1974. Internal surges in coastal water. *J. Geophys. Res.*, 79: 4523–4526.

[20] Gargett, A. 1980. Turbulence measurements through a train of breaking internal waves in Knight Inlet, B.C. in *Fjord Oceanography,* ed. H. J. Freeland *et al.* NATO Conference Series IV, Marine Science, Vol. 4.

[21] Farmer, D. M. & Armi, L. 1986. Maximal two-layer flow over a sill and through the combination of a sill and contraction with barotropic flow. *J. Fluid Mech.*, 164: 53–76.

[22] Gross, M. G. 1972. *Oceanography: A view of the Earth.* Prentice Hall, N. J.

8

Industrial problems with gravity currents: oceanography.

8.1 POWER STATION EFFLUENTS

Fossil fuel and nuclear power stations raise steam to drive turbines. Their efficiency is less than about 40 per cent and the balance of the heat produced is rejected to the condenser cooling water; if it is impracticable to make use of this heat, it must be dissipated in the environment.

At coastal stations the obvious method of cooling the power stations is by direct, or once-through, cooling. Water is pumped from the sea, through the station condensers where it takes up latent heat from the condensing turbine steam, and is returned to the sea from the cooling water outlet.

The cooling water requirement per GW of generating capacity is about $30 \, \text{m}^3 \, \text{s}^{-1}$ or $50 \, \text{m}^3 \, \text{s}^{-1}$ for a nuclear plant [1]. The freshwater flow in the River Thames at Teddington Lock averages $75 \, \text{m}^3 \, \text{s}^{-1}$ so that the cooling water from a 2 GW power station can be compared with a fairly large natural river.

This cooling water from the sea is heated up about 10°C and discharged back again. It is important to know where the discharged water goes and how quickly it cools down, both for ecological reasons (the discharge might affect marine life) and for station efficiency; recirculation, i.e. the reuse of already warmed water, lowers efficiency.

When the cooling water leaves the outlet, it tends to rise to the surface and spread out horizontally. Very close to the outlet, in the so-called "near-field", the discharge is dominated by the momentum of the flow.

A few hundred metres further away, in the "mid-field", the discharge behaves much like a gravity current. That is to say, sharp boundaries are maintained between the outflow and denser surrounding water, and these are maintained even when the flow is disturbed by variations of depth or turbulence in the surrounding water.

In the "far-field", turbulence processes dominate and no sharp horizontal boundary can be seen between the flow and the ambient water.

Fig. 8.1 sums up what is thought to happen to a radially symmetrical rising jet in

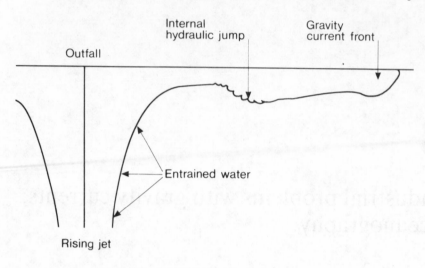

Fig. 8.1 — Rising jet over deep still water.

deep still water, showing the near- and mid-fields. The rising jet entrains surrounding water and spreads horizontally. If this flow is still supercritical, then an internal hydraulic jump is expected, leading to the gravity current head shown at the front of the flow. The typical value of reduced gravity, g', at the outflow is 0.025 m s^{-2} for a 10° rise in temperature, leading, for a 1 m deep current, for example, to a spreading velocity of 6 cm s^{-1} (or 200 m h^{-1}), of the order of the spreading observed.

British coastal waters are tidal and the spread of the cooling water plume is very sensitive to the ambient current. Extensive research on cooling water in tidal flows has been carried out at Sizewell power station on the east coast of England [2].

Fig. 8.2 gives an interpretation of such a flow during the flood tide, with vertical scales exaggerated. There is considerable mixing in A, the initial stage during the rise from the outfall ports. In B, the stable gravity current stage, there is less vertical mixing. In the far-field, C, the buoyancy forces no longer dominate and heat is transported vertically by turbulent processes. In Fig. 8.2, as in most models of this process, the plume is shown to thicken with increasing distance away from the outfall due to a combination of interfacial shear entrainment and mixing caused by the turbulent surroundings. Some recent models have a different viewpoint of turbulent mixing [3] predicting that turbulence in the ambient water could have the effect, not of diffusing the warm water, but of restraining it in a well-defined layer, while the sub-plume layer gradually warms up fairly uniformly through its layer. Experiments described in Chapter 14 give some idea of how these two processes can come about.

8.2   OIL SLICKS

The spreading of oil over the surface of the sea is an environmental process in which gravity currents play an important part. As oil production and oil transport over the sea continue to increase, the public have become familiar with the frequent damage caused by oil spillage. A few spills of the order of 100 000 tons of crude oil have

a. *(plan)* Buoyant plume at flood tide

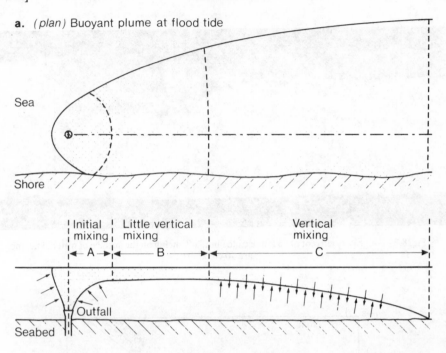

b. *(section)* Mixing of plume water parallel to shore

Fig. 8.2 — Plan and elevation of cooling water plume. A, initial stage. B, gravity current stage. C, far field.

focussed world-wide attention on the problem and there has been much anxiety over the problems of where the oil goes and of the area covered.

Firstly, winds and ocean currents move the oil mass as a whole. Secondly, since the oil has a density 10 to 20 per cent less than water, gravity-controlled spreading must be very significant in the behaviour of the oil.

The initial growth of the slick is similar to the early stages of any gravity current immediately after release, and the behaviour which follows has been investigated both theoretically and in the laboratory. Fig. 8.3, taken during a laboratory experiment, shows the front of a gravity current of immiscible fluid (kerosene) flowing above water. The flow here is in the gravity–inertial regime and shows a turbulent zone behind the deepened head of the current.

The flow which follows the initial collapse can be divided into three regimes, gravity–inertial, gravity–viscous and viscous–surface [4,5]. Fig. 8.4 shows a time scale for these regimes in the development of a typical oil slick, the depth of which will be of the order of 1 centimetre.

In the gravity–inertial regime, the viscous drag at the oil–water interface is negligible. In the gravity–viscous regime, the thickness is small compared with the previous regime and the drag of the water on the slick becomes dominant. After a certain time oil spilled on the sea ceases to spread. The spreading process ends in a surface-tension regime.

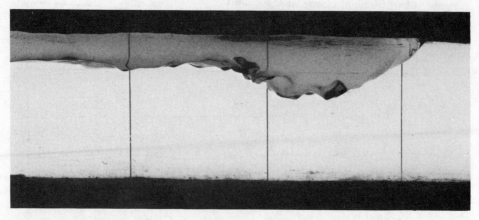

Fig. 8.3 — A gravity current of an immiscible fluid, advancing on the surface of water. The lines
are 10 cm apart.

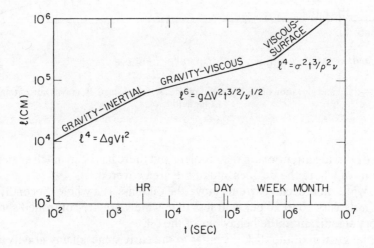

Fig. 8.4 — Combined spreading laws. (Courtesy of D. P. Hoult).

Time scales may be:

gravity–inertial	1000 s
gravity–viscous	up to 1 week
viscous–surface	1 week to 1 month

8.2.1 Containment of oil slicks

Floating booms have been used to try to contain oil on a water surface, but their
effectiveness is limited by currents, winds and waves. Although they are unsatisfac-
tory in the open sea they have provided effective containment in sheltered harbours,
rivers and estuaries.

Even when protected from wave action, an oil boom can fail by the mechanism of oil underflow when oil is swept under the boom by currents. The forms of oil slicks arrested by booms have been investigated experimentally [6] and it has been shown that oil will tend to build up against such a barrier. A slick will propagate upstream as increasing quantities of oil are contained and three possible outcomes are illustrated in Fig. 8.5.

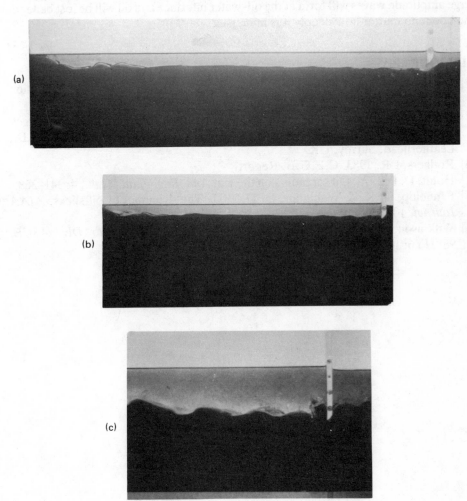

Fig. 8.5 — Laboratory experiments. (a) Oil slick contained by barrier. (b) Oil escaping beyond barrier of insufficient depth. (c) Oil slick with unstable interface. (Courtesy of D. L. Wilkinson).

In case (a) the oil–water interface of the slick is stable and, if the upstream front is far enough away, interfacial disturbances there will die away and a stable wedge can be contained by the barrier.

The second type of slick in (b) may occur when the barrier is not deep enough to retain the oil in an otherwise stable slick. The oil will continually drain beneath the barrier, and in time nearly all the oil will be lost.

The third type of slick (c) has no equilibrium thickness and the waves and mixing at the interface make retention of the slick impossible. The dynamic instability which makes containment impossible occurs when the Froude number of the flow upstream of the slick exceeds a critical value. In flows with Froude number above this value large amplitude waves will form at the oil–water interface and oil will be lost beneath the boom no matter how deeply it is immersed.

BIBLIOGRAPHY

[1] Macqueen, J. F. 1978. Background water temperatures and power station discharges. *Advances in Water Resources,* 1: 195–203.
[2] Ewing, D. J. F. 1982. *The spreading-out of cooling-water discharged from direct-cooled power stations.* Central Research Laboratories Report, C.E.G.B., Leatherhead, Surrey, UK.
[3] Rodgers, I.R. 1983. C.E.G.B. Report.
[4] Hoult, D. P. 1972. Oil spreading on the sea. *Ann. Rev. Fluid Mech.,* 4: 341–368.
[5] Fannelop, T. K. & Waldman, G. D. 1971. The dynamics of oil slicks. *AIAA Journal,* 10: 506–510.
[6] Wilkinson, D. L. 1972. Dynamics of contained oil slicks. *J. Hydr. Div., ASCE,* 98, HY6; 1013–1030.

9

Avalanches

9.1 TYPES OF AVALANCHE

Many different cases of mass transport occurring under gravity have been studied in geological science. A well-known example of these is the snow avalanche, which has already been mentioned in Chapter 1. There are several other types of avalanche; one of the most significant of these is an avalanche of debris or mud which can be even more destructive and dangerous.

The whole range of debris flows represents a set of processes lying between the drier forms of mass transport and liquid stream flow. At the latter end of this range, mud flows can occur. These have a high content of water and may be derived from hill slopes which have disintegrated with catastrophic effect.

Volcanic eruptions are responsible for several types of gravity-controlled flows including debris and mud flows. Two of these are uniquely volcanic: pyroclastic flows and streams of basaltic lava. These will be dealt with in detail in Chapter 10.

9.2 SNOW AVALANCHES

People living in alpine regions are exposed in winter and spring to avalanches of snow, and research on the dynamics and prevention of avalanches is carried out by most of the mountainous nations. There are many historical records of snow avalanches, among others of their creation by the passage of Hannibal's men and animals across the Alps in 218 B.C.

An early illustration of an avalanche is shown in Fig. 9.1, a wood engraving by H. Schäufelein in the "Theurdanck" of 1517. This pictures an avalanche as a series of large snowballs; as late as the end of the nineteenth century people continued to imagine avalanches as giant snowballs which increased by accretion of the underlying snow as they bounded down the mountain side.

Since about 1900 organised scientific research on avalanches has developed in many countries, notably in Switzerland, France, Japan, Canada and USA, and its importance is still growing. The areas of practical importance in this study can be divided into three: the starting zone, the avalanche track and the run-out area.

Fig. 9.1 — Sixteenth century illustration of a snow avalanche.

9.2.1 Starting Zone

First, in the "starting zone", there is the problem of the release mechanism. Snow is not a simple material and the mechanism of the initial fracture is almost impossibly complex theoretically. In spite of the large body of information on the properties of snow that can be found in a recent review [1], prediction of avalanche releases remains essentially empirical. The snow cover may break away to a depth of the order of 1 metre to generate an avalanche and this may be caused by changes of the structure of the snow. Other possible causes of avalanche triggering are ski loads, falling cornices and earthquakes. They are also initiated artificially.

9.2.2 Types of snow avalanche

In the second stage, the mature avalanche travels down the slope as a gravity current. Snow avalanches may be crudely classified into two types, *flow avalanches* and *airborne powder snow avalanches*. Most avalanches do not belong to either of these limiting cases, but are rather a mixture of the two. The two limiting cases are shown in Figs 9.2 and 9.3 respectively. Wet-snow avalanches form a class of flow avalanches and have some interesting features in common with debris flows and mud avalanches, to be dealt with in Section 9.4

Flow avalanches have a typical velocity of up to 60 m s^{-1}, and a flow height of 5 to 10 m. Powder-snow avalanches move at velocities over 100 m s^{-1} and can be over 100 metres high. The graph in Fig. 9.4 shows a set of observations from field work of the

Fig. 9.2 — Flow avalanche.

velocity of both powder-snow avalanches and flow avalanches [2], plotted against the depth of the flow.

A flow avalanche initially tends to slide like a rigid body but rapidly breaks up into smaller particles and turns into a granular material flow. The "fluidisation" mechanism for such a flow depends on the interchange of energy of rapidly colliding small particles, and will be discussed again in the following section on dry rock avalanches.

An airborne powder-snow avalanche is essentially a turbidity current flowing down an incline. The behaviour of slope gravity currents and turbidity currents in the laboratory will be described in Chapters 11 and 16. Additional work to clarify the entrainment process in powder snow avalanches has been carried out, together with numerical models of the flows [3].

Mechanisms are illustrated in Fig. 9.5 by which the suspension of snow particles is

Fig. 9.3 — Powder snow avalanche.

built up, to form a cloud 100 m deep. The turbulent flow over the ground may result in entrainment of further fresh snow from beneath and more air will be entrained above the head.

It is generally agreed that powder snow avalanches develop from flow avalanches, but the mechanism by which the transition from one regime to the other takes place is not clearly understood. A trailing cloud often develops behind a flow-avalanche which may increase in volume and actually overtake the dense snow body if the track is long. In these cases a powder-snow avalanche may form very rapidly, as the air entrainment grows explosively.

9.2.3 Runout zone

It is important in practice to know when an avalanche will come to a stop and it has been shown that obstacles can reduce the runout distance [4,5]. In contrast to flow avalanches, powder snow avalanches behave like Newtonian fluids and flow around an obstacle. Generally a fast moving dry-snow avalanche is accompanied by a cloud riding above it; when the avalanche hits a barrier the dense portion is stopped and the cloud behaves like a fluid flowing round an obstacle. A low barrier will usually have the effect of accelerating the flow just above it, and damage can be increased. Fig. 9.6 shows the results of a related laboratory experiment in which the flow is shown passing a barrier. Although there are strong eddies behind the obstruction, the results show an appreciable protected zone.

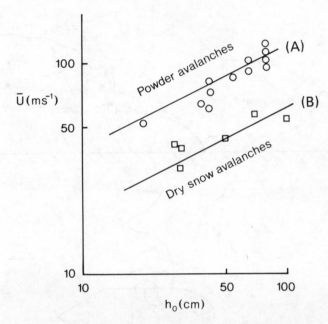

Fig. 9.4 — Observed avalanche velocity as a function of snow depth. (A) Powder avalanches. (B) Dry snow avalanches. (Courtesy of E. J. Hopfinger).

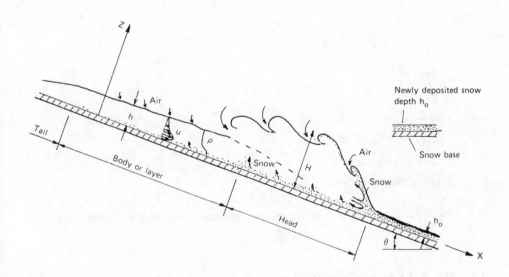

Fig. 9.5 — Parts of an avalanche. A large cloud of snow dust lies above the turbulent zone of dense snow.

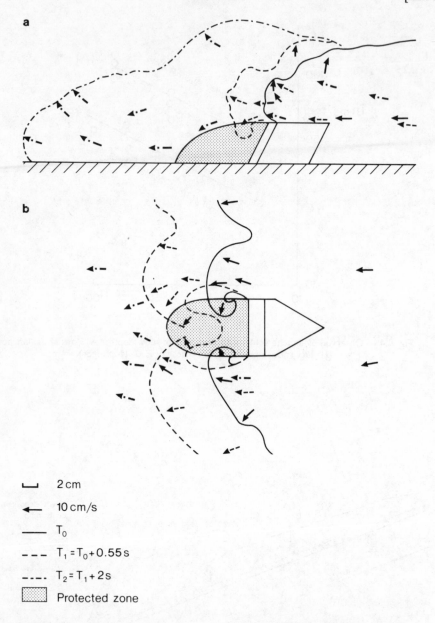

Fig. 9.6 — The flow of an avalanche past an obstruction. (Courtesy of E. J. Hopfinger).

9.3 ROCK AVALANCHES (DRY)

Small rockfalls can be derived from soil creep or freeze–thaw, but the larger ones are usually caused by earthquakes. After a short transitional stage the rockfall may develop into an avalanche of rocks of assorted sizes which forms a fluidised flow.

Rockfalls of relatively small volume occur frequently in mountainous regions, and the inhabitants have learned to live beyond their range. Occasionally, however, rockfalls of unusually large volume occur. The material from these travels long distances, perhaps many kilometres beyond what is expected of a mass of solids sliding down an inclined plane with normal coefficient of friction.

In 1881 the Alpine village of Elm in Switzerland was virtually wiped out by one of these enormous rockfalls which flowed 2 km down the valley. This disaster made geologists and engineers aware that large masses of rock debris may sometimes behave like a fluid with a low internal resistance. Since 1881 many other instances of fluidised rock avalanches have been found, both contemporary and historic. Such mobile debris streams, sometimes known as "Sturzstroms", have been extremely costly in human lives and, despite many investigations, no universal explanation has been agreed for the extremely long travel distances of rock avalanches.

Fig. 9.7 shows the travel distance of Sturzstrom against the volume of rock

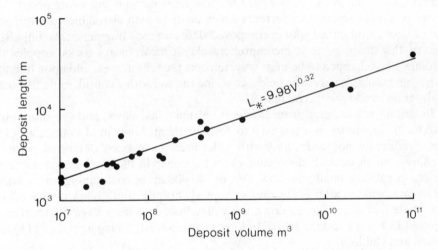

Fig. 9.7 — Deposit length of rock avalanches plotted against deposit volume. (After Davies, 1982 [6]).

deposited [6]. This plot is for volumes greater than 10^7 m^3, the range associated with unexpectedly long travel distance. These results strongly suggest that the deposit extent of a rock avalanche depends mainly on its volume, and not the height through which it has fallen. This suggests a fluid-like spreading rather than a sliding mechanism.

Possible explanations in terms of proposed fluidisation mechanisms include:

(1) Air fluidisation. This cannot explain all the observations, because similar flows have been inferred from deposits on the airless surfaces of the Moon and Mars.
(2) Basal melting theory. Much reduced friction could be accounted for if there were sufficient melting of the rock at the points of contact.
(3) Acoustic fluidisation. Fluidisation by sound energy is believed to be possible

through the medium of strong acoustic waves, or "noise" within the avalanche mass. It is thought that sufficiently large amounts of sound energy can be maintained in large enough flows [7].

(4) Mechanical fluidisation. The essence of mechanical fluidisation is that a high energy input to a mass of granular material causes high impulsive contact pressures, due to high shearing rates, as it moves over the ground beneath. As a result the particles become randomly separated and the mass dilates; the internal resistance to the deformation of the material as a whole is thus reduced. This reduction of internal resistance has been shown in laboratory experiments.

Mechanical fluidisation seems well able to explain the behaviour of rock avalanches, but it is possible that all these mechanisms play a part.

9.4   DEBRIS AND MUD AVALANCHES (WET)

After a visit to the Alps in the 1780's the naturalist de Saussure wrote about "the danger of being surprised by torrents which descend with incredible velocity. . . . a kind of liquid mud mixed with decomposed slate and rock fragments; the impulsive force of this dense paste is incomprehensible; it incorporates rocks, topples the buildings which happen to be in its way, uproots the tallest trees, and upon bursting forth from ravines ravages the fields, covering the soil with a considerable thickness of silt, gravel and rock fragments".

Indonesia has suffered from frequent volcanic mud flows, and the Indonesian word for them, *lahars*, is often used to describe them. Many mud avalanches of all sizes, frequently not connected with volcanoes, have been described since de Saussure; one particularly distressing event happened in 1966 in Wales as the result of a comparatively small mud flow. About 140 000 m^3 of coal-mining waste, which had become sodden after heavy rain, collapsed abruptly, and a black tongue of coal dust, shale and water flowed down the valley floor into the village of Aberfan. It engulfed 18 houses and the school before it came to rest, killing a total of 144 men, women and children.

The speed at which these flows travel depends on many things, including the steepness of the slope and the mass and viscosity of the slurry. It can be very fast indeed, of the order of 90 km h^{-1}. Values of the kinematic viscosity for mud flows have been estimated as between 2×10^3 and 6×10^3 SI units, as compared with the value of 10^{-2} for water at 20°C.

Alluvial fans consist of deposits of coarse and fine debris which are brought down from mountain ranges and spread out on the plain. Only occasional reports have been given of the behaviour of the actual mud flows which are their origin. A graphic description has been given of the appearance of a debris and mud flow as it descended from the White Mountains, on the border of California and Nevada, contributing to an alluvial fan at the base of the mountains. The following events were reconstructed from interviews with two stockmen whose ranches were at the edge of the fan [8].

(a) Two hours after a heavy thunderstorm in the mountains on the afternoon of 26 July 1952, loud rumbling and roaring noises were heard emanating from the lower canyons of the affected drainages.

(b) About 30 minutes later, masses of debris were noticed advancing downslope on the upper parts of the fans. The leading edge appeared to be a low wall of boulders and thick mud, without visible water.

(c) The debris appeared to be advancing in a series of waves or surges, each wave overtaking and submerging the preceding one.

(d) The flows were accompanied by noise likened to "the sound of thousands of freight cars bumping together simultaneously".

(e) At the lower part of the canyon the material moved "about as fast as man can dog-trot", perhaps 400 or 500 feet per minute. It was 1 to 2 feet thick.

The photographs in Fig. 9.8 show smaller-scale debris flows as developed on the cone of sediment rejected from a gravel washing pit. The flow in (a) shows the formation of a levee at the sides of the upper slope and the spreading out in the plain as the gradient decreases. Fig. 9.8 (b) shows clearly the overhanging snouts of the flow which is only a few centimetres thick.

Dam-bursts have often resulted in catastrophic mudflows; one such disaster occurred in the Italian Dolomites in 1985 when 260 people were killed. A dam burst and released a wall of sludge, said to be 20 m high, which overwhelmed the small village of Stava. The fluidised material consisted of the earth and rubble of which the dam had been built and large quantities of deposit previously lying at the bed of the lake.

The dangers of mud avalanches are increasing for several reasons. Not only are more people living in their path, but they are removing natural defences. As populations increase and pressures on land grow, developing countries are removing stabilising vegetation. The risks are being increased by permitting overgrazing, by destroying the previous land cover for crops and by cutting timber without control. Air pollution is a significant factor in killing trees. It has justly been said that "debris flows may be natural, but they are not averse to a helping hand".

9.4.1 Instabilities in debris and mud flows

We have seen how debris avalanches can flow like a liquid and that the ability to carry big boulders is the crucial characteristic of true debris flow. Field observations from a variety of sources show that debris flows in gently graded channels tend to a fairly uniform thickness, but a mode of instability appears under some conditions of speed, density and slope, these flows showing a pulsing nature with a series of large waves or surges.

The loss of strength which accompanies the remoulding of muddy sediments and introduces a mud flow may, in some cases, be reversible. Materials which behave in this way are called *thixotropic*; some examples are emulsion paint, strong suspensions of cornflower or custard powder, and many aqueous clay mineral dispersions. A thixotropic fluid has been used to demonstrate a mode of instability in laboratory suspension flow; Fig. 9.9 shows a very strong suspension of custard powder in water moving down a slope of 20° in a tank of fresh water.

Mud flows generated by heavy rainfall are rather common in the mountainous region of south-west China and frequently cause catastrophic damage to the local villages and fields. Some of the largest and most hazardous mud flows occur in the

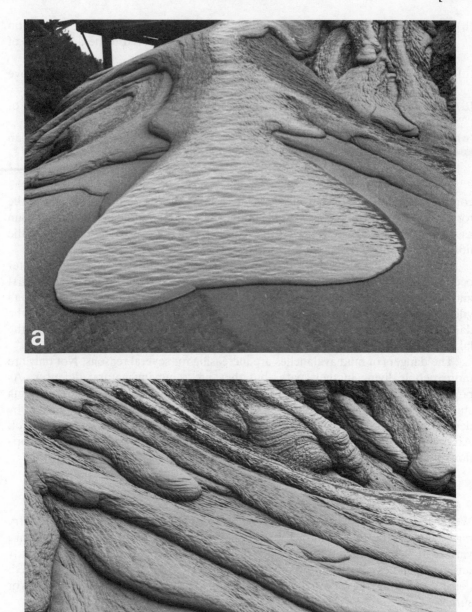

Fig. 9.8 — Debris flow of muddy sand developed from the cone of sediment from a gravel washing plant. (a) An active flow. (b) The thinner "dead" flow. (Courtesy of J. R. L. Allen).

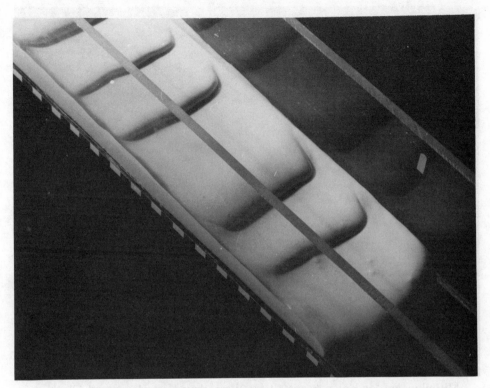

Fig. 9.9 — Instability on the surface of a laboratory model of a debris flow. The material used is a very thick suspension of custard powder.

Jiang-jia Ravine, a district which has been carefully studied by scientists from Sichuan Province [9].

One important feature which has been discovered from observations at the research station in the Ravine has been named the "bursting" process. This mudflow usually bursts in the form of successive waves, examples of which are illustrated in Fig. 9.10. The wave trains occur from time to time and there is little or no flow between successive groups of waves. The flow may cease altogether for a period of 20 to 30 minutes. This period of silence ends with a distant sound like thunder marking the next convulsion, and the surges of the mudflow rush down one after another until the episode is ended.

This kind of behaviour depends on the physical nature of the material and the rheological features of the fluid flow ; the field work suggests that the transport of large solids causes the pulsing.

BIBLIOGRAPHY

[1] Salm, B. 1982. Mechanical problems of snow. *Rev. Geophys. Space Phys.*, 20: 1–19.

[2] Hopfinger, E. J. 1983. Snow avalanche motion and related phenomena. *Ann. Rev. Fluid Mech.*, 15: 47–76.

Fig. 9.10 — Advancing surge of mud flow in the Jiang-jia Ravine. (Courtesy of Sichuan Geographical Institute).

[3] Scheiwiller, T. 1986. *Dynamics of powder-snow avalanches*. Report No. 81, Mitteilungen der Versuchsanstalt für Wasserbau Hydrologie und Glaziologie, Zürich. 115 pp.

[4] Salm, B. 1966. Contribution to avalanche dynamics. *Int. Symp. on Scientific Aspects of Snow and Ice Avalanches, Davos, Switzerland. AIHS Publ.*, 69: 199–214.

[5] Hopfinger, E. J. & Tochon-Danguy, J-C. 1977. A model study of powder snow avalanches. *J. Glaciol.*, 19: 343–356.

[6] Davies, T. R. H. 1982. Spreading of rock avalanche debris by mechanical fluidization. *Rock Mechanics*, 15: 9–24.

[7] Melosh, H. J. 1980. Acoustic fluidization: a new geologic process. *J. Geophys. Res.* 84: 7513–7520.

[8] Beaty, C. B. 1963. Origin of alluvial fans, White Mountains, California and Nevada. *Ann. Ass. Am. Geog.*, 53: 516–535.

[9] Li Jian, Yuan Jianmo, Bi Cheng & Luo Defu 1983. The main features of the mud flow in Jiang-jia Ravine. *Z. Geomorph. N.F.*, 27: 325–341.

10

Volcanic gravity currents

Many types of gravity current are associated with volcanic eruptions. The emergence of material from the magma, the molten rock beneath the ground, forms currents the nature of which depends on the viscosity of the magma and on the quantity of dissolved gas contained under pressure.

At one end of the range are flows of basaltic lava, which is comparatively fluid molten rock at temperatures of 1200°C which emerges quietly and flows as a glowing stream down the mountainside. At the other extreme are laval flows of very viscous molten rock rich in silica. Such flows form steep-sided plugs and domes, and can take months to move a few hundred metres.

If the magma contains large quantities of gas and has a very high temperature, the nature of the eruption may be quite different. A vertical eruption column of glowing material may be projected hundreds of metres up into the air; it then falls back under gravity and forms a gravity current of fluidised material called a pyroclastic (literally "fiery broken") flow. Such a current is sometimes called a *nuée ardente* or fiery cloud. Pyroclastic flows from the largest eruptions may still have velocities of up to 100 m^{-1} at distances of tens of kilometres from the source [1], so they form a serious hazard to people near volcanoes.

10.1 BASALTIC LAVA STREAMS

Basalt is the commonest kind of lava and when erupted is a runny liquid, glowing reddish-yellow at a temperature of about 1200°C. Lavas are comparatively harmless, as one can see them coming and usually get out of the way quickly enough. Fig. 10.1 shows a thin lava flow on the south–west slope of Surtsey, Iceland, in 1963. It can be seen flowing down to the sea from the lava pool higher up the mountain.

As it cools, the lava gradually stiffens and develops a chilled skin which affects the nature of the flow. This skin becomes wrinkled by the flow movement, giving a characteristic ropey appearance. As the material continues to stiffen, banks of solid material, *levees*, form at the sides. The whole flow may even end up confined to the inside of a solidified lava tunnel.

Fig. 10.1 — Lava flow on Surtsey. The lava flows from a lava-pool down to the sea. (Photograph by Solarfilma, Reykjavik).

In the later stages, solid lumps of lava pile up in front of the advancing nose which rolls forward and engulfs the material, rolling it up like the tracks of a caterpillar tractor. Later still there is no longer any fluid flow over the mass of rubble which is then continually piled up and pushed forward by the fluid behind it until at last the whole system comes to rest.

10.1.1 Very viscous lava flows

Silica-rich rock may still flow as a very viscous fluid. Such flows move at the speed of a glacier and can take months to cover a few hundred metres.

A good example of a lava dome formed by very viscous, slow moving lava was studied in 1979 in the crater of Soufrière Volcano, St. Vincent [2]. The extruded lava formed a dome 130 metres high and 870 metres in diameter. The map of the crater in Fig. 10.2 shows how, in a period of three months, the leading edge of this actively growing viscous gravity current moved about 200 metres. Views of the Soufrière extrusion, taken on 4 August 1979, are shown in Fig. 10.3. A theory for such viscous flows was tested in the laboratory (see Chapter 15) and gave good results when applied to the Soufrière dome.

10.2 PYROCLASTIC FLOWS: NUÉES ARDENTES

At the other end of the mobility scale from the flows of basaltic lava we have pyroclastic flows of which the "nuée ardente" is an example. Some of the very different forms taken by pyroclastic flows can be associated with the early stages of

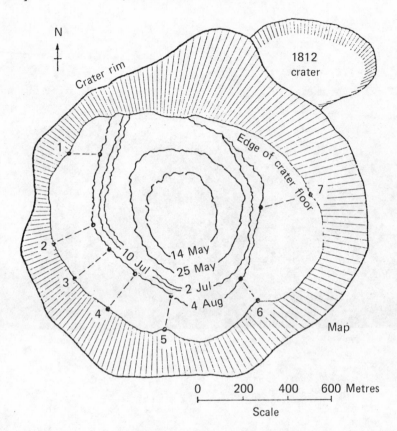

Fig. 10.2 — Map of the Soufrière crater, showing outlines of the lava extrusion at different times. (Courtesy of H. E. Huppert).

their formation. Examples of different generating processes are shown in Fig. 10.4, where (a) shows the eruption of material vertically into the air; this collapses to the ground and initiates a fiery gravity current down the slope; (b) shows the pressure is not so great and the fluid just boils over onto the slope; and in (c) the flow may be very vigorous, as explosions direct the volcanic material sideways. This occurred at Mount St. Helens.

The first pyroclastic flows to be studied scientifically were from Mount Pelée on Martinique in the Caribbean. The town of St. Pierre was overwhelmed by a vast nuée ardente in 1902 and all but two of its inhabitants were killed [1]. After some more eruptions of the same volcano in 1929 Perret studied hundreds of nuées ardentes and took a remarkable series of photographs [3]. Perret was fascinated by the appearance of the advancing front, and wrote ". . . convolutions grew out of a rolling mass of incandescent material, advancing with an indescribably curious rolling and puffing movement which at the immediate front takes the form of forward-springing jets, suggesting charging lions" — a graphic description of the many gravity current fronts illustrated in this book.

The comparatively recent photograph of a nuée ardente in Fig. 10.5 was taken at

Fig. 10.3 — Views of the Soufrière lava extrusion on 4 August 1979. (Courtesy of H. E. Huppert).

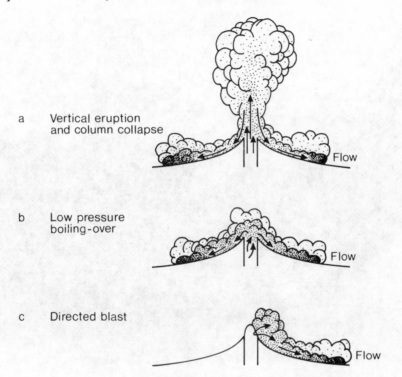

a   Vertical eruption
    and column collapse

Flow

b   Low pressure
    boiling-over

Flow

c   Directed blast

Flow

Fig. 10.4 — Some ways that pyroclastic flows can originate. (After Macdonald, 1972, *Volcanoes*, Prentice-Hall, Inc.).

Mount St. Helens in 1980. Although little of the complicated internal structure can be deduced from this view, it does show the raised nose and lobe-and-cleft structure of a typical gravity current.

As mentioned above, one method in which pyroclastic flows are generated is by gravitational collapse of material ejected from a volcano where the magma is especially viscous, and contains large amounts of dissolved gas. A good picture of the process involved is of a fizzy drink being shot from a bottle which has been shaken. When the stopper is removed, fluid is blown with great force through the opening, and as the pressure is reduced large quantities of gas are released, forming a turbulent jet of material and breaking up the fluid into small drops.

Two stages in the life-history of a small nuée ardente are shown in the photographs of Fig. 10.6. The first photograph was taken only 8.2 seconds after the start of an eruption at Ngauruhoe, New Zealand, and it shows that in this short time the ejected material had already formed a large cloud. There are slight signs beneath this cloud of descending material. The second view was taken 90 seconds after the start, by which time a nuée ardente had formed from the material descending at the peak and its leading edge had already moved as far as the base of the mountain.

Mechanisms of fluidisation believed to be responsible for the rapid flow of other types of avalanche have already been discussed, in relation both to laboratory

Fig. 10.5 — Formation of a pyroclastic gravity current on Mt. St. Helens, 7 August 1980.
(Photograph by Harry J. Glicken, courtesy of US Geological Survey).

experiments and to avalanches of snow and rock. An additional process which may
be important in the mechanics of these pyroclastic flows is the release of large
quantities of gas which may come out of solution from the hot material. Some
measurements of the actual speeds of nuées ardentes were made from an aeroplane
flying over the eruptions of Mount Augustine Volcano, Alaska, in 1976 [4]. Along
the initial 1:3 slope from 1200 m to 240 m an average speed of 50 m s^{-1} was attained.
The speed reduced to 21 m s^{-1} down the subsequent 1:7 gradient, and the flow finally
entered the sea, 6 km from the volcanic cone, after a time of 6.7 minutes.

From measurements of the deposits left behind nuées ardentes, deductions have
been made about their internal progress. The transported mass may become more
dense near the ground, and a buoyant cloud may separate [5, 6].

A nuée ardente was mentioned above which entered the sea, and the entrance of
hot pyroclastic flows into water must be a frequent event on island volcanoes. There
is evidence that these flows can move under water without losing their essential
character and their deposits have been traced to over 13 km offshore at a water depth
of 1800 m [7]. Only flows denser than water are capable of a smooth transition into
water, and conditions favourable for the passage into deep water include steep slopes
and large flow rates. Laboratory experiments [8] suggest that the shearing which
occurs at the upper surface of the flow front generates a turbulent zone of mixing
which may continue as a turbidity current after the debris flow has come to rest.

Fig. 10.6 — Formation of nuée ardente on Ngauruhoe, New Zealand. Times after start of eruption: (A) 8.2 seconds, (B) 90 seconds. (Courtesy of G. T. Hancox, NZ Geological Survey).

10.2.1 Base surges

When water has access to the rising magma column in a volcano, steam-rich eruptions may boil outwards in dense clouds. Fig. 10.7 shows a sketch of the resulting ring-like collar, often named a "base surge" [9].

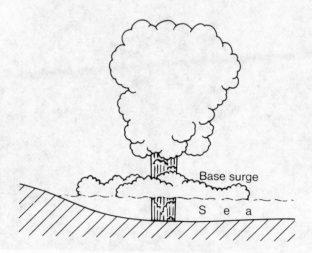

Fig. 10.7 — A base surge, a developing ring-like collar of cloud.

Eruptions that produce base surges involve release of large volumes of steam capable of supporting or fluidising many of the particles from the central eruption column. The flowing mixture has a mobility similar to that of a nuée ardente, but its particles are wet instead of hot. This spreads radially as a gravity current of considerable power.

The deposits laid down by base surges on steep slopes sometimes contain U-shaped furrows which are believed to be related to the lobe and cleft structure observed at gravity current heads [10]. Lobes and clefts can certainly be seen at the heads of base surges, and it is thought that the increased speed of flow through the lobes is responsible for the formation of these patterns.

10.3 MUD FLOWS

Mud flows are common on many volcanoes but their importance is not always appreciated by the casually interested as they tend to get lost in press reports in vague description of "eruptions". Volcanic eruptions may trigger off mud flows both indirectly and directly. An eruption ejecting ash high into the atmosphere may propagate a tropical rainstorm, which in turn may initiate the kind of cold mud flow described in Chapter 9. Much larger mud flows can result when an eruption blasts through a crater lake, expelling the water and a great deal of volcanic debris at the same time. The lake water may be boiling hot, so that the mud flow down the mountain is a lethal mixture of scalding water, mud and boulders.

Fig. 10.8 shows a phenomenon which is common in Iceland, where several active volcanoes lie beneath large ice sheets. Enormous quantities of water are melted beneath the ice-cap during a volcanic eruption and burst out, running down the mountain side. The resulting flow of slurry is known as a *jökulhlaup*.

Mud flows, or lahars, have claimed more lives worldwide over the past 200 years than any other type of avalanche. One of the worst disasters was caused by the volcanic eruption of Nevado del Ruis in Colombia on 13 November 1985, which killed 25 000 people. This large eruption produced pyroclastic flows and surges which generated four major lahars; these stripped vegetation and soil from the canyon walls to heights of over 50 m and travelled a total distance of 60 km at a mean speed of 10 m s^{-1} [11]. In the town of Armero the lahars overtopped the banks of the river channel and followed the old river course directly into the town, where most of the casualties were caused.

Analysis has been carried out of the remains of the debris flow in the canyon 2.5 km from Armero where the lahars stripped vegetation and soil from the canyon walls and deposited a layer of mud. Fig. 10.9 shows the superelevation of the flow along the right bank as it rushed round the corner. This superelevation has been used [11] to calculate directly the speed of the debris flows:

$$w = (g \cos A \tan B.R)^{\frac{1}{2}}$$

where g is the acceleration due to gravity, A the slope of the stream bed, B the slope of the banked flow surface (deduced from differential levels of scour on the opposite sides of the river), and R is the radius of curvature of the bend. Around this bend, g is 9.8 m s^{-1}, B is 7.5° and R is 110 m, giving a mean front velocity of 11.9 m s^{-1}. This flow was about 45 m deep at mid-channel and had a cross-sectional area of nearly 4000 m^2.

One group of investigators [11] has pointed out that the large volume of glacial ice on the summit of the mountain, the abundance of debris around the summit and the precipitous gorges leading down to river valleys collectively represent ideal conditions for the formation and long-distance movement of lahars.

10.4 ERUPTION OF MOUNT ST. HELENS

Descriptions of the explosive eruptions at Mount St. Helens, Washington, in 1980 illustrate well some of the processes mentioned above, namely pyroclastic flows, rock avalanches and mud flows [12,13].

After two months of minor earthquake activity, in the morning of 18 May 1980, there was an immense detonation and the entire north side of the volcano began to separate from the main mass. A diagrammatic illustration is shown in Fig. 10.10.

The detachment of the giant rockslide triggered explosions releasing a laterally directed pyroclastic surge from the mountainside. This material swept forward at over 90 m s^{-1}, overriding the rock avalanche, and devastated a sector reaching 20 km from the mountain summit. In an inner zone of 10 km radius virtually everything was destroyed, and beyond this radius all trees were blown down.

The progress of the pyroclastic flow seemed almost independent of the contours of the ground, but the advance of the rock and mud flows was somewhat different.

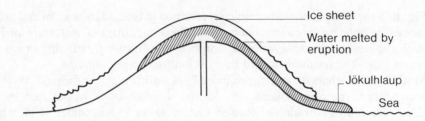

Fig. 10.8 — A jökulhlaup, a flow from a sub-glacial eruption.

Fig. 10.9 — Canyon looking upstream, 2.5 km from Armero, showing area stripped by lahar. A man (arrowed) shows the scale. The superelevation of the flow along the north (right) band can be seen. (Courtesy of *Nature*, photograph by Donald R. Lowe, Department of Geology, Louisiana State University).

The avalanching rock from the north flank acquired melted snow and ice blocks from the former glaciers of the volcano. Travelling at 80 km per hour this debris flow was diverted to the left by a ridge, displaced the water of Spirit Lake and flowed down the valley of the Toutle River. For over 18 km this lahar swept down the valley, causing damage, including destruction of bridges and houses. The deposit left in the upper reaches of the Toutle River had an average thickness of 66 m.

The total number of people killed was less than 100, a comparatively small

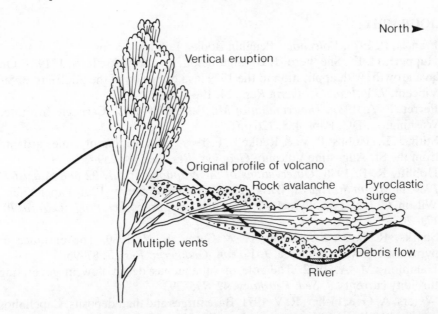

Fig. 10.10 — Section through Mt. St. Helens, showing the effect of the sideways blast in the eruption of 18 May 1980.

number for such a vigorous eruption because of evacuation and restricted access to the area.

10.5 THE LAKE NYOS GAS DISASTER

On 21 August 1986 a large volume of toxic gas was released from beneath and within Lake Nyos in the mountains of Cameroon. A vast gravity current consisting of an aerosol of toxic gases and water droplets swept down the valleys to the north, killing 1700 people.

The conclusion of a group of British investigators after visiting the site [14] are as follows. The waters of Lake Nyos were already saturated with CO_2 of volcanic origin when a pulse of volcanic gas, mainly of CO_2 but with some H_2S, was released in the lake above a volcanic vent. The rising stream of bubbles brought up bottom water containing CO_2 dissolved under pressure. As the pressure was released the gas came out of solution extremely vigorously, as when released in a fizzy drink, and increased the flow of water to the surface.

At the surface the release of gas transformed the water into a fine mist and this aerosol of water and heavy gases swept down the valleys to the north of the lake. It was estimated that about 200 000 tonnes of water were lost from the lake, together with 6000 tonnes of gas which at atmospheric temperature and pressure would have a volume of about 3 million cubic metres.

Whilst this disaster is included in the Section on volcanoes, the apparently stealthy approach of the aerosol current could classify it with the gas spills described in Section 6.2.

BIBLIOGRAPHY

[1] Francis, P. 1976. *Volcanoes*. Penguin Books, London. 368 pp.

[2] Huppert, H. E., Shepherd, J. B., Sigurdsson, H. & Sparks, R. S. J. 1982. On lava growth, with application to the 1979 lava extrusion of the Soufrière of St. Vincent. *J. Volcan. Geotherm Res.*, 14: 199–222.

[3] Perret, F. A. 1937. *The eruption of Mt. Pelée 1929–1932*. Carnegie Institute, Washington, DC, Publ. 458. 126 pp.

[4] Stith, J. L., Hobbs. P. V. & Radke L. F. 1977. Observations of a nuée ardente from the St. Augustine Volcano. *Geophys. Res. Lett.*, 4: 259–262.

[5] Hoblitt, R. P. 1986. *Observations of the eruptions of July 22 and August 7, 1980, at Mount St. Helens, Washington*. U.S. Geol. Survey Prof. Paper 1335.

[6] Wilson, C. J. N. 1986. Pyroclastic flows and ignimbrites. *Sci. Prog., Oxford.* 70: 171–207.

[7] Sparks, R. S. J., Sigurdsson, H, & Carey, S. N. 1980. The entrance of pyroclastic flows into the sea. *J. Volcan. Geotherm. Res.*, 7: 87–96.

[8] Hampton, M. A. 1972. The role of subaqueous debris flow in generating turbidity currents. *J. Sed. Petrology*, 42: 775–793.

[9] Waters, A. G. & Fisher, R. V. 1971. Base surges and their deposits: Capelinhos and Taal Volcanoes. *J. Geophys. Res.*, 76: 5596–5614.

[10] Fisher, R. V. 1977. Erosion by volcanic base-surge density currents: U-shaped channels. *Geol. Soc. Amer. Bull.*, 88: 1287–1297.

[11] Lowe, D. R., Williams, S. N., Leigh, H., Conner, C. B., Gemmell, J. B. & Stoiber, R. E. 1986. Lahars initiated by the 13 November 1985 eruption of Nevado del Ruiz, Colombia. *Nature*, 324: 51–53.

[12] Moore, J. C. & Rice, C. J. 1984. *Explosive Eruptions of Mount Saint Helens*. in: Explosive Volcanism: Inception, Evolution, and Hazards. National Academy of Sciences, Washington, D.C., 133–142.

[13] Christiansen, R. L., 1980. Eruption of Mt. St. Helens: Volcanology. *Nature*, 285: 531–533.

[14] Freeth, S. J. & Lay, R. L. F. 1986. The Lake Nyos disaster. *Nature*, 325, 104–105.

11

The anatomy of a gravity current

11.1 THE FRONT

The leading edge of a gravity current forms a typical frontal zone, i.e. although intense mixing is present, a sharp dividing line is maintained between the two fluids. A characteristic "head" which is deeper than the following flow is usually formed at the front. This raised head is a zone of breaking waves and intense mixing and plays an important part in the control of the current which follows. In a gravity current moving horizontally the head remains quasi-steady but in one flowing down an incline the relative size of the head increases with the angle of the slope.

A front advancing over a rigid flat surface usually has a foremost point or "nose" which is raised above the following flow. Both the raised nose and the zone of intense mixing can be seen clearly in the photograph of the atmospheric gravity current, or "haboob", illustrated in Fig. 1.1.

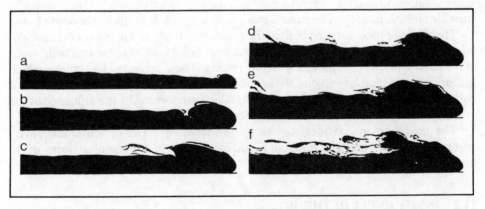

Fig. 11.1 — Shadow pictures showing profiles of the head of a gravity current. The temperature difference increases from very small in (a) to 35° in (f). (From Schmidt, 1911 [1]).

It is not possible to give a unique shape for the outline of the head of a gravity current, as even in flows into calm surroundings the excess head height above the following flow varies with the fraction of the total depth of the fluid occupied by the current. Apart from viscous effects, the head profile is also strongly modified by opposing or following ambient flows, and by turbulence in the surroundings.

The influence of viscosity on the behaviour of gravity currents has been examined in the laboratory, and it has been shown that viscosity affects the profile of the head and the rate of advance of the current. The results of some early experiments [1] showed this well, as illustrated in Fig. 11.1. A series of gravity currents of increasing temperature difference were used to model the advance of a cold squall in the atmosphere. These shadow-pictures show gravity currents with a temperature difference increasing from a very small value in (a) to 35° in (f), corresponding to a density difference of about 1%. As the temperature difference increases, and with it the speed of the gravity current, the shape of the head of the current changes. In the flow (a) at small temperature difference, viscous forces predominate over the forces due to buoyancy; as a result the head is small and there is very little mixing apparent. As the density difference increases the flow approaches the limiting profile in (f) in which eddies are formed and can be seen streaming back in the upper surface of the head.

The values of Reynolds number, UH/v, in the experiments shown in Fig. 11.1 increase from less than 10 in (a) to greater than 1000 in (f). The final view (f) shows a profile which has been shown [2] to be independent of Reynolds number, that is, to be typical of all flows with Re greater than 1000. In all these flows a turbulent mixing pattern is seen streaming back in the upper surface of the head, and the height of the foremost point of the nose is about a tenth of the total height of the head. The Reynolds number of a typical thunderstorm outflow in the atmosphere is about 1 000 000, and many field observations have shown the structure to be similar to that seen in the laboratory with Re greater than 1000.

The features of the head of a gravity current can be seen in the view of a laboratory experiment shown in Fig. 11.2. This shows the frontal region of a 1% saline solution advancing along the floor of a parallel-sided channel of fresh water 30 cm wide and 20 cm deep. The front is made visible by adding milk to the dense fluid.

The mixing processes seen both in the laboratory flows and in many environmental examples of gravity currents are complicated. Fig. 11.3 shows the two main types of instability which are responsible for the mixing. These consist of (A) billows which roll up in the region of velocity shear above the front of the dense fluid and (B) a complex shifting pattern of lobes and clefts which are formed by the influence of the ground on the lower part of the leading edge.

The next section considers the simplified front of a two-dimensional gravity current advancing along a horizontal surface, firstly in the simplest form of an inviscid flow with no mixing.

11.2 INVISCID-FLUID THEORY

Inviscid-fluid theory has been applied to study aspects of a steady gravity current; in particular it shows the role of wave-breaking and the associated energy losses. In an inviscid fluid, in which viscous forces are completely absent, the frictional effect of the ground disappears but the instability still remains that leads to the formation of billows in the head. Such theory would be expected to give a useful approximation to the behaviour of the front of a gravity current, at least for flows with Reynolds

Fig. 11.2 — The frontal region of a gravity current of salt water advancing along the floor of a freshwater tank.

number above a certain value. Nevertheless, some of the details must be different from flows in real fluids, in which frictional forces certainly play a part. Such a theoretical "inviscid current" cannot be realised in nature, but can be approached experimentally in certain ways.

The front of a frictionless gravity current has been analysed [3] in terms of a "cavity flow" displacing a fluid beneath it. If one end is removed from a long closed water-filled channel, the water will begin to run out at the lower level, being replaced by air flowing in above it. Fig. 11.4 shows the form of the air intrusion. The velocity U_2 is relative to the leading edge of the cavity.

Two equations involving the velocity and depth h_2 of the flowing layer can be obtained from continuity of mass and by the use of Bernouilli's equation applied along the interface. (Bernouilli's equation expresses the constancy of total energy along a streamline, where the sum of the three types of energy in the flow — pressure, potential and kinetic — remains constant.) Being frictionless the "flow force" (total pressure force plus momentum flux per unit span) is also conserved, resulting in a solution

$$h_2 = H/2$$

Thus the only steady energy-conserving flow (implied by the use of Bernoulli's equation) is one in which the advancing layer fills half the channel. Flows in which

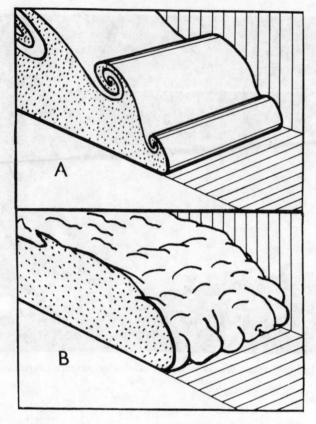

Fig. 11.3 — Two forms of instability at the front of a gravity current moving along the ground:
(A) billows; (B) lobes and clefts.

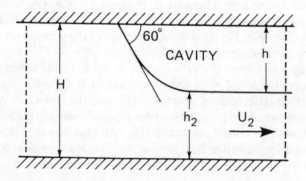

Fig. 11.4 — Theoretical treatment of a gravity current head. The fluid is displaced as it passes
the cavity.

$h>H/2$ are not possible, and if $h<H/2$, as is found in most practical situations, the loss of energy at the front exceeds that available by wave radiation so that "breaking" must occur.

It can also be shown that the fractional depth $h/H = \phi$, say, plays an important part. The Froude number, $U/(g'h)^{\frac{1}{2}}$, based on the velocity U of the front, is $2^{-\frac{1}{2}}$ at $\phi = 1/5$, and equals about 1 at $\phi = 1/2$, increasing to $2^{\frac{1}{2}}$ as ϕ tends to 0.

In subsequent treatment of the velocity of the head under "deep" water (i.e. ϕ tends to 0), if the pressure downstream in the wake is taken to be hydrostatic so that the dynamic pressure $\frac{1}{2}\rho U^2$ at the stagnation point at the front boundary equals the difference between the hydrostatic pressure at the boundary far upstream and downstream, it follows that

$$\tfrac{1}{2}\rho U^2 = g(\rho_1 - \rho_2)h$$

or

$$U = 2^{\frac{1}{2}}(g'h)^{\frac{1}{2}}$$

The same result had been previously obtained [4] by applying the Bernoulli condition along the interface, but this is invalid in deep water since the interface must be dissipative and Bernoulli's equation cannot be applied.

The approximate shape of the interface has also been calculated, and the slope at the stagnation point on the ground can be shown to be 60°.

11.3   EFFECT OF MIXING ON GRAVITY CURRENTS

Most gravity currents in the environment have turbulent mixing at the front, but there may be only small frictional effects if there is no rigid boundary. An example of this is a large-scale gravity current of fresh water moving above salt water, such as a river flowing out over the sea. In this case there is considerable mixing at the interface between the fresh and salt water, and this plays an important part in the dynamics of the flow.

A semi-empirical analysis of this type of front has been carried out as a first step towards realistic modelling of the head of such gravity currents [5]. In this analysis the flow relative to the head is divided into three regions, as shown in Fig. 11.5. This represents a gravity current of dense fluid, miscible with the surrounding fluid, moving along a plane surface with no friction.

The bottom region, depth h_4, represents the flow of dense unmixed fluid into the gravity current front. The top region contains only the unmixed less dense fluid, depth h_2, passing above the head. The section, h_3, between these layers is the mixing region. This region has non-uniform velocity and concentration profiles, determined by experiment.

The apparatus shown in Fig. 11.6 was used to examine the properties of the mixing at the head of a gravity current with effectively no friction at the floor. It consists of a parallel-sided channel of perspex with water pumped through it at a steady rate. The floor of the left part of the channel consists of a flexible conveyor-

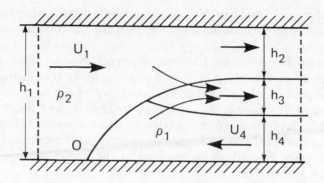

Fig. 11.5—The flow relative to an inviscid gravity current with mixing. The bottom region is the flow of dense unmixed fluid into the front; above it is the mixed region of the collapsed billows.

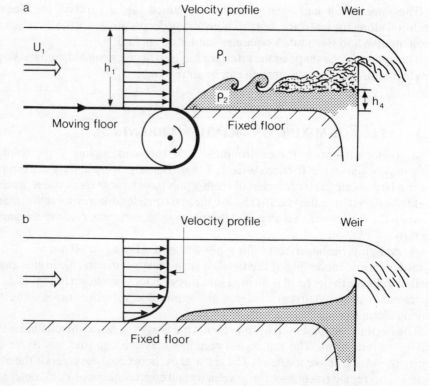

Fig. 11.6 — Apparatus used to maintain the head of a gravity current in a steady state. The moving floor is used to modify the velocity profile in the approaching flow.

belt which can be moved at the same speed as the water. A steady supply of dense fluid is introduced through a meter from beneath at the right end of the tank.

For a fixed input of dense fluid of a given density it is possible to adjust the flow-and-floor speed to bring the dense fluid to rest just downstream of the conveyor-belt, as shown in the top of Fig. 11.6, forming a front at rest at the beginning of the fixed floor section.

It is important to realise the difference between this front and the familiar "arrested saline wedge" of dense fluid brought to rest by a simple opposing flow along the ground. The wedge has little or no mixing and will be dealt with in more detail in Section 11.5.1. As shown in Fig. 11.6, in case (a) with the floor moving at the same speed as the water flow, the front is brought to rest by a relative flow having a constant velocity profile, U. The effect is therefore the same as a front moving at velocity U into a calm fluid. However, in case (b) conditions are quite different, the wedge being brought to rest by a flow whose velocity is reduced by friction to zero close to the ground.

The nature of the imposed velocity profile has a critical effect on the instability and hence the mixing developing along the foremost interface. The result of the profile in (b) is to reduce the sharpness of the velocity profile in the flow along the interface; as a result there is an almost complete absence of instability at the front of the arrested wedge.

When the front is held stationary on the fixed floor just downstream of the end of the conveyor-belt section it appears as shown in Fig. 11.7. This photograph was

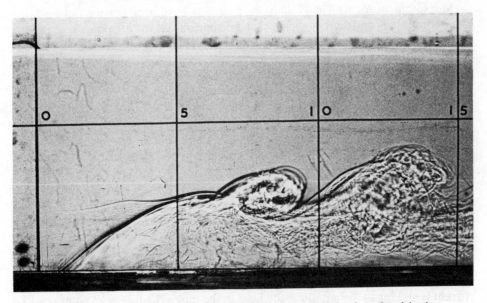

Fig. 11.7 — Shadowgraph of the head of a gravity current head produced in the apparatus of Fig. 11.6. The lobe-and-cleft instability has disappeared and the billows are two-dimensional.

obtained by a simple "shadowgraph" technique, which is a simplified form of the "schlieren" process, using deflection of the light by density gradients in the fluid under examination. A slide-projector is used at a distance of about 5 metres (or more) to project a shadow of the flow on a transparent screen. It can be seen that the foremost point of the dense fluid is on the ground and it is found that any lobe-and-cleft instability (illustrated in Fig. 11.3) has completely disappeared. The mixing zone above the head can be seen to contain clear two-dimensional billows.

To determine the velocity of the front theoretically, values of the Froude number $U/(g'h_4)^{\frac{1}{2}}$ were calculated using the continuity and Bernoulli equations applied along the floor to the foremost stagnation point, O. This Froude number varied both with the fractional depth h_4/h_1 and also with $q = g'Q/U^3$, the non-dimensional mixing rate, where Q is the actual rate of mixing. Values of q needed to close the equations could be either measured experimentally or else deduced from billow properties.

Experimental measurements of the mixing rate were obtained in the apparatus of Fig. 11.6 by simply monitoring the rate of input of dense fluid needed to maintain the steady state of the front.

This apparatus also made it possible to investigate the properties of the billows forming at the front, and it was found that they had the properties of Kelvin–Helmholtz billows. This type of billow is associated with instability formed at the interface between two fluids of different density, moving relative to each other. The instability can occur for certain ranges of values of the velocity change U per distance h and density difference given by reduced gravity g'. The relevant dimensionless number here is the Richardson number Ri, which may be put in the form $(g'h)/U^2$. Kelvin–Helmholtz instability usually occurs for a value of Ri less than $\frac{1}{4}$.

Measurements of Kelvin–Helmholtz instability have been carried out for large ranges of velocity and density shear [6] and it appears that the breakdown size of the billows at a gravity current head agrees with that of K–H billows forming at a very low Richardson number (i.e. a very sharp interface, as exists at the foremost part of a gravity current). The height of a gravity current head is identical with the breakdown size of the K–H billows which are being formed there.

Using either of these methods of obtaining the non-dimensional mixing rate, the Froude number was found theoretically, and confirmed by experiment, to be close to 1 for fractional depth h_4/h_1 about one fifth of the total depth. For smaller fractional depths, down to about 0.05, the value of the Froude number rose to approximately 2.

11.4   EFFECT OF FRICTION

The outline of the front of a gravity current advancing along the ground has appeared in several of the environmental photographs, and shows a rather complicated system of billows, and also lobes and clefts. The shadowgraph in Fig. 11.8 gives us a look inside such a front in the laboratory. An almost unmixed dense saline flow can be seen entering the water tank on the right at the bottom, and the front of the gravity current is moving to the left along the floor of the tank. The foremost point is raised from the ground, and scattered Kelvin–Helmholtz billows can be seen forming above and collapsing to the right of the head.

A schematic diagram in Fig. 11.9 shows a simplified two-dimensional flow pattern relative to such a gravity current head. In such a flow, with no-slip conditions at the

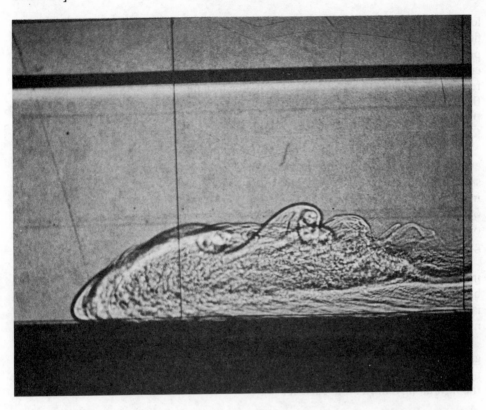

Fig. 11.8 — Shadowgraph of laboratory gravity current on horizontal floor. Salt solution enters right.

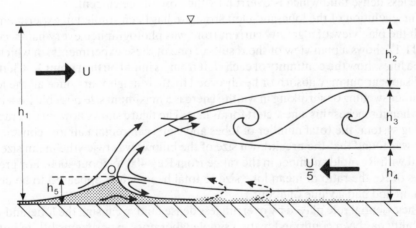

Fig. 11.9 — The flow relative to the head of a gravity current advancing along a horizontal surface. The foremost point O is raised above the floor, and the shaded fluid passes beneath the head. There is a small reverse circulation close to the ground.

lower boundary due to friction at the stationary ground, the lowest streamlines in the flow relative to the head must be towards the rear. This means that the stagnation point O must be raised a small distance above the floor, and in addition to the circulation in the upper part of the head there must be a smaller circulation in the reverse sense close to the ground. The fluid which has been shaded in the diagram is that which is destined to pass underneath the nose of the current. This is less dense then the fluid above it and therefore is unstable. In the process of its ascent and forward movement it is responsible for the non-steady lobe and cleft structure.

In this type of flow along a solid boundary the billows which form at the leading edge are broken up in a complicated three-dimensional form. Nevertheless by using special lighting it is still possible to distinguish perfectly formed Kelvin–Helmholtz instabilties [7]. An example appears in Fig. 11.10 which shows three successive views of a cross-section of a gravity current head taken at intervals of $\frac{1}{4}$ second. This was obtained by using a narrow sheet of light which illuminated a cross-section of a gravity current marked by a fluorescent dye.

The billows appear to be the main process in which the surrounding fluid is mixed into a gravity current. The fluxes in and out of the frontal region of a gravity current have been measured using hot-wire probes on a carriage moving with the front [8]. It is found that the mass flux of light liquid into the front is of the order of 0.15 times the mass flux of the current itself.

To summarise, the three graphs in Fig. 11.11 show the variation of velocity of a gravity current with fractional depth. The speed is non-dimensionised as the Froude number, $U/(g'/h_4)^{\frac{1}{2}}$; (1) is the inviscid flow, (2) the inviscid flow with mixing, and (3) the flow along a horizontal surface.

11.4.1 Lobes and clefts

The complicated shifting pattern of lobes and clefts at the head of a gravity current moving along a horizontal surface is believed to be caused by gravitational instability of the less dense fluid which is overrun by the nose of the current.

The evolution of the lobe-and-cleft structure has been studied in experiments in which the plan view of a gravity current front was photographed every half-second. Fig. 11.12 shows a plan view of the results of one of these experiments, in which the dotted lines show the continuity of each cleft from its initial birth at point X. Clefts do not disappear but may absorb or be absorbed by their neighbours since all the lobes are either swelling or shrinking in width. There is a maximum size possible for a lobe, and when it reaches this a new cleft forms in it. The figure shows how, in this rapidly shifting system, the total number of lobes and clefts can remain almost constant.

It was found that the breakdown size of the billows was twice the mean size and varied with Reynolds number in the range from Re=400 to about 4000. For greater values of Re the ratio of mean lobe size to total head height appeared to be nearly constant and had a value of about $\frac{1}{4}$.

The essential role played by overrun less dense fluid in causing the lobe-and-cleft instability has been confirmed by two simple laboratory experiments [9], in both of which it was found possible to suppress the instability.

In the first experiment, the overrunning of light fluid was prevented by laying down a thin layer of dense fluid ahead of the current. This was deep enough to ensure

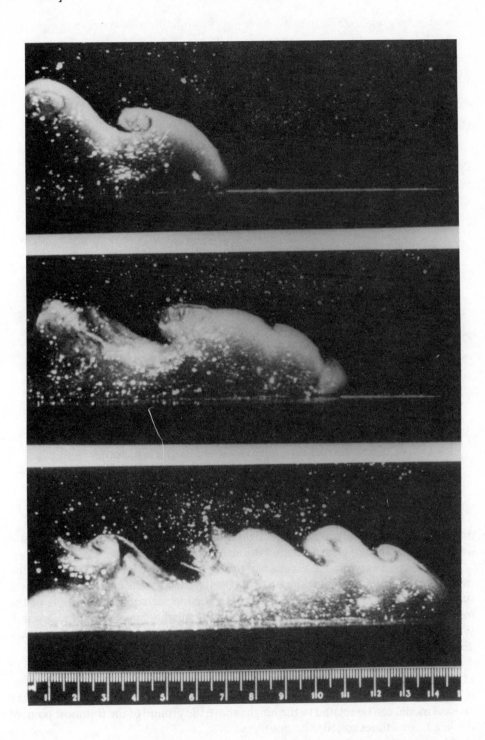

Fig. 11.10 — Three succesive views of a gravity current head at intervals of $\frac{1}{4}$ second, in which some clear two-dimensional billows have been detected.

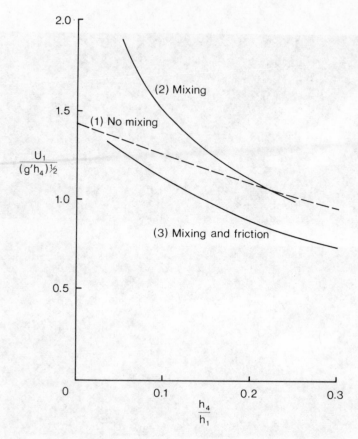

Fig. 11.11 — The speed of the head of a gravity current advancing in fluid of different depths. The speed is expressed as the Froude number and the depth as the fraction of the current of the total depth. Curve (1) is the inviscid current with no mixing, (2) is an inviscid current with mixing and (3) is a mixing current moving along the ground.

that all the overrun fluid was no longer lighter than the gravity current fluid; the result was the appearance of clear two-dimensional billows.

In the second experiment a section of the floor was moved in the direction of the current, but only beneath the dense current itself. Ahead of the current the floor was stationary as usual. As the speed of the floor-section increased, the profile of the front became flatter, but also the foremost point steadily approached the ground. When the foremost point was actually on the ground the flow duly became two-dimensional and clear K–H billows appeared.

11.4.2 Height of the foremost point, or nose

If we now include the effect of friction on the boundary, the main difference from the inviscid model can be related to the height above the ground of the foremost point of the head, sometimes called the "nose".

Measurements of the height of the nose of a gravity current advancing along the ground into calm surroundings have been made in the laboratory and in the

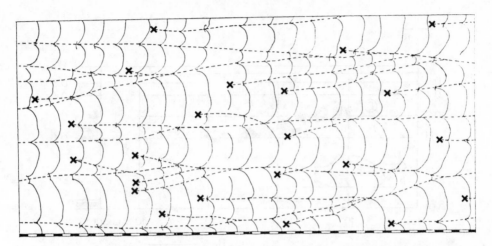

Fig. 11.12 — Successive views of the leading edge of a gravity current seen from above at intervals of ½ second, showing the evolution of lobes and clefts. X marks the points where new clefts appear. The scale is in centimetres, $\Delta\rho/\rho$ is 1% and total water depth is 24 cm.

atmosphere and some results are shown in Fig. 11.13 [10]. Laboratory experimental results are shown for values of the Reynolds number $U_1(h_3+h_4)/\nu$, from 10 to 10^5, and atmospheric results are at about Re=10^8. The results show a height of the foremost point of about ⅛ the head height, apparently independent of Reynolds number of values greater than 10^3.

11.5 HEAD AND TAIL AMBIENT FLOWS

The profile of the head of a gravity current and its rate of advance are very sensitive to any opposing or following flow in the environment. These head- and tail-flow effects on the front of a gravity current moving along a horizontal surface have been examined in a water channel with a moving floor of the type already described in Section 11.4 [5].

Using this apparatus, illustrated in Fig. 11.14, head flows, or "head winds", are simulated by bringing the front to rest on the moving floor with an opposing flow, U_1, greater than that of the floor, U_0. Tail winds are obtained by moving the floor faster than the opposing flow.

The experiments show that with a head wind the head profile is longer and the nose height lower than in the calm case already illustrated. With a tail wind the head is shorter and the nose height is larger. Fig. 11.15 shows these three different forms.

A head or tail wind is found to change the speed of advance along the ground by about three-fifths of the applied wind. Values of the overtaking speed relative to the front, U_4, were compared with U_1, the flow relative to the ground, and the values of U_4/U_1 were all found to be close to 0.15, independent of the values of the head or tail flows [10].

The value of the mixing rate, $R_L=g'H_3/(\Delta U)^2$, was deduced to be 0.5 in all head and tail wind cases. This is greater than the values measured in the two-dimensional

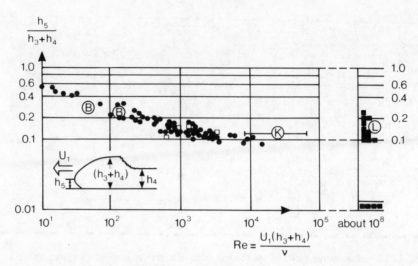

Fig. 11.13 — Values of the height of the foremost point of a gravity current head measured for flows at different Reynolds numbers. $Re=U_1(h_3+h_4)/\nu$. The figures on the right are from atmospheric observations.

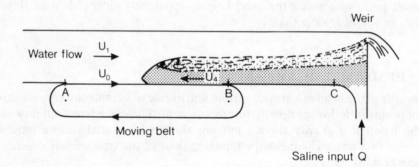

Fig. 11.14 — Schematic diagram of water channel with moving floor used to investigate the dynamics of gravity current head. The speed of the water flow, the speed of the moving floor and the input of dense fluid can all be controlled.

(inviscid) case, but close to that already found in the calm case along a horizontal floor. Thus it appears that the shape of the head and mixed region adjusts to maintain this almost constant layer Richardson number.

11.5.1 Arrested wedges

In many rivers, as the tide level increases, a gravity current of dense salt water moves upstream along the river bed. Eventually the front is brought to rest and for some time an arrested saline wedge exists.

The laboratory experiments shown above in Fig. 11.6 displayed an arrested

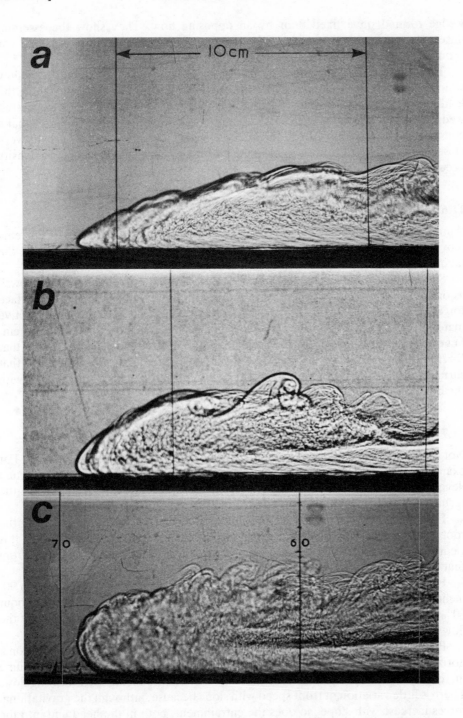

Fig. 11.15 — Shadowgraphs show the effect of head and tail winds on the form of a gravity current head. In the experimental tank the vertical lines are 10 cm apart. (a) Flow speed greater than floor speed (head wind); (b) flow speed equal to floor speed (calm); (c) flow speed less than floor speed (tail wind).

wedge formed on a fixed floor by an opposing flow. They show the essential difference between this wedge and a gravity current moving into still fluid (or brought to rest by opposing flow and floor).

The difference was explained by the velocity profile near the ground in the flow which is meeting the wedge. This has the effect of reducing the instability along the leading edge of the dense fluid. In fact the mixing along the top surface of an arrested wedge may be almost non-existent. Laboratory experiments [11] have shown that a considerable proportion of the interface of such a saline underflow can be very closely approximated to a straight line, and that no appreciable change in density occurs during the length of the underflow.

11.6   FRONT MOVING ALONG A FREE SURFACE

Plumes of fresh water flowing above salt sea-water are examples of gravity currents moving along a surface free to the atmosphere. Modelling these flows in the laboratory needs care with the selection of scale.

Small-scale laboratory buoyant gravity currents, sometimes called "overflows" as opposed to "underflows" along the ground, can be strongly affected by surface forces. These are caused by surface tension and by films resulting from surface-active materials in the water. These effects may be large and it is difficult to reduce them, but experiments [12] have shown that surface tension effects are small provided that $\Sigma = \Delta\sigma/g'h$ is smaller than 0.01 (where $\Delta\sigma$ is the surface tension difference), and that surface film effects are not important when the current depth is large compared with the boundary layer depth.

11.7   GRAVITY CURRENTS ON SLOPES

Not all gravity currents in the environment flow along horizontal surfaces. For example the sloping surface of a mountain plays a vital part in the formation and development of a snow avalanche. A slope is also needed for the formation of the turbidity currents which form in the ocean at the edge of the continental shelf.

From direct measurements of the behaviour of buoyant methane gas along the roof of a passage of a mine [13] and from later laboratory experiments [14,15], it is clear that the motion of a gravity current down a slope is appreciably different from that along a horizontal surface.

For a given rate of flow of dense material on slopes from a few degrees to 90° inclination there is a nearly constant front velocity. This velocity is found to be about 60% of the mean velocity of the following flow, or in other words about 40% of the flow arriving at the front is mixed into the head.

Fig. 11.16 shows three different forms of the head of a gravity current: first on a horizontal surface and then down slopes of 5° and 20° respectively. The head volume increases, both by direct entrainment and also by addition from the following flow. There is little variation of front speed with slope because, although the gravitational forces increase with slope, so does the entrainment, both in the head and into the flow behind it.

The amount of direct entrainment increases with slope, leading to one-tenth of the head-growth at 10° and about two-thirds at 90°. With a steady buoyancy flux Q,

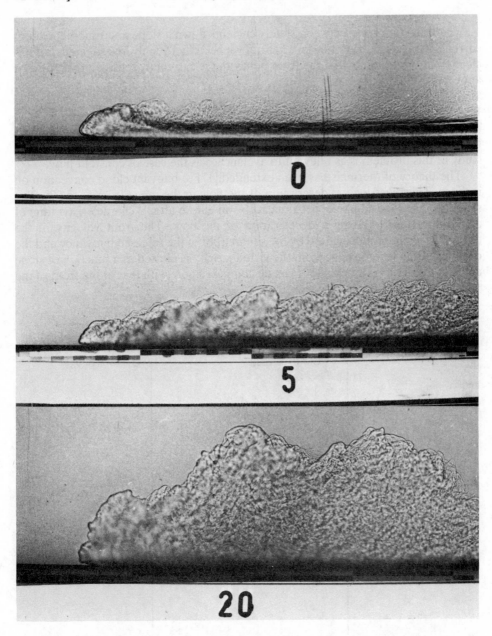

Fig. 11.16 — Gravity currents flowing down slopes, of 0°, 5° and 20°. The entrainment increases with slope, both into the head and into the flow behind it. (Britter & Linden, 1980 [16]).

experiments [16] showed a constant head velocity for slopes greater than 5°, and from 5° to 90° the front velocity was given by

$$U/(g'Q)^{\frac{1}{3}} = 1.5 \pm 0.2$$

The transition zone between steady currents down a slope and time-dependent currents along a horizontal surface has not yet been thoroughly investigated, but it is clear from experiments that the front behaviour is extremely sensitive to slope as it approaches the horizontal.

11.7.1 Motion of an "inclined thermal"

The release of a finite volume of dense fluid down a slope and the resulting "inclined thermal" of "fixed-volume cloud" have practical applications. For example, the study of this problem is as relevant to avalanches as the case of constant flow.

The theory of thermals has been extended [17] to buoyant cloud convection on inclines. This gives a good description of the flow in the range from 5° to 90°. The shape of the cloud can be approximated by an ellipse after a characteristic time of acceleration, followed by a deceleration of the flow. The front velocity in the decelerating phase, normalised by the square root of the released buoyancy and the distance of the leading edge from the virtual origin, is plotted as a function of slope angle in Fig. 11.17. Its behaviour can be compared here with that of the front of the

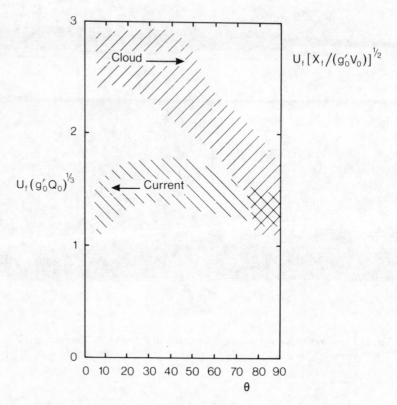

Fig. 11.17 — Comparison of the speed down a slope, between a continuous gravity current and a "cloud" or "thermal".

steady current, discussed above. The experimental scatter in results is large, due to the variability in structure of the front, but it can be seen that the dependence with angle of slope is more pronounced in the thermal cloud than in the steady current counterpart. A velocity maximum in the "thermal" is again observed at slope angles between 10° and 40°.

BIBLIOGRAPHY

[1] Schmidt, W. 1911. Zur Mechanik der Boen. *Z. Meteorol.* 28: 355–362.

[2] Simpson, J. E. & Britter, R. E. 1979. The dynamics of the head of a gravity current advancing over a horizontal surface. *J. Fluid Mech.*, 94: 477–495.

[3] Benjamin, T. B. 1968. Gravity currents and related phenomena. *J. Fluid Mech.*, 31: 209–243.

[4] von Karman, T. 1940. The engineer grapples with non-linear problems. *Bull. Am. Math. Soc.*, 46: 615–683.

[5] Britter, R. E. & Simpson, J. E. 1978. Experiments on the dynamics of a gravity current head. *J. Fluid Mech.*, 88: 223–240.

[6] Thorpe, S. A. 1973. Experiments on instability and turbulence in a stratified shear flow. *J. Fluid Mech.*, 61: 731–751.

[7] Simpson, J. E. 1969. A comparison between laboratory and atmospheric density currents. *Quart. J. Roy. Met. Soc.*, 95: 758–765.

[8] Winant, C. D. & Bratkovich, A. 1977. Structure and mixing within the frontal region of a density current. *6th Aust. Hydraul. & Fluid Mech. Conf.*, Adelaide, pp. 9–12.

[9] Simpson, J. E. 1972. Effects of the lower boundary on the head of a gravity current. *J. Fluid Mech.*, 53: 759–768.

[10] Simpson, J. E. & Britter, R. E. 1980. A laboratory model of an atmospheric mesofront. *Quart. J. Roy. Met. Soc.*, 106: 485–500.

[11] Riddell, J. C. 1970. Arrested saline wedge. *Houille Blanche*, 4/1970: 317–330.

[12] Thomas, N. H. & Simpson, J. E. 1983. *The motion of a gravity current along a free surface.* SEGB report.

[13] Georgeson, E. M. H. 1942. The free streaming of gases in sloping channels. *Proc. R. Soc., London, Ser. A*, 180: 484–493.

[14] Middleton, G. V. 1966. Experiments on density and turbidity currents. I. Motion of the head. *Can. J. Earth Sci.*, 3: 523–546.

[15] Hopfinger, E. J. & Tochon-Danguy, J. C. 1977. A model study of powder-snow avalanches. *Glaciology*, 19: 343–356.

[16] Britter, R. E. & Linden, P. F. 1980. The motion of the front of a gravity current travelling down an incline. *J. Fluid Mech.*, 99: 531–543.

[17] Beghin, P., Hopfinger, E. J. & Britter, R. E. 1981. Gravitational convection from instantaneous sources on inclined boundaries. *J. Fluid Mech.*, 107: 407–422.

12

Spread of dense fluid

The previous chapter considered in detail the advance of the front of a gravity current when its depth and density difference have already been established. This front plays an important part in the dynamics of a gravity current, but to deal with the flow as a whole, one needs to know how the fluid is being supplied and the form of the surroundings into which it is allowed to spread.

12.1 LOCK-EXCHANGE FLOWS

Some of the earliest measurements of gravity currents were made in navigation canals near the coast where exchanges of fresh and salt water always take place whenever a lock gate is opened.

To set up an experimental system, the channel is temporarily divided into two sections by a thin vertical barrier. Fresh water is run into one segment and salt water into the other, and the levels are made equal. As soon as the barrier is raised, the dense fluid starts to collapse and counter currents begin to flow in opposite directions. These consist of a gravity current of less dense fluid moving along the surface and a dense current flowing beneath it in the other direction.

Many experiments of this nature have been carried out both in canals [1] and on a smaller scale in the laboratory [2]. Lock-exchange flows in a channel of uniform rectangular cross-section will be considered first. The most notable feature of these flows is the uniformity of the front velocity. The front continues to advance at a constant speed, provided that the two parts of the channel are both very long and that the flow does not reach the range of viscous control (see Chapter 15).

Several authors have given the same analysis, in which the extension of a rectangular block as in Fig. 12.1(a) is assumed as a first approximation to lock exchange flow. By equating decrease of potential energy with increase of kinetic energy, the following result is obtained:

$$U_0/(g'H)^{\frac{1}{2}} = 0.5$$

This forecast is very close to the value of 0.46 for the velocity found in an extended series of experiments in very large tanks [3].

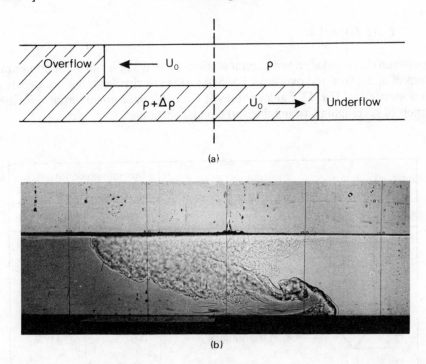

Fig. 12.1 — Developing lock-exchange flow. (a) Inviscid, non-mixing theoretical model. (b) Shadowgraph of flow in laboratory experiment.

The photograph in Fig. 12.1(b) is of a developing lock-exchange flow in the laboratory. This shows some features which are different from the simple theoretical model. Each flow has a head slightly deeper than the following flow and the flow has a slight slope behind it.

The front along the free surface advances slightly faster than that along the lower boundary and a value of 0.59 has been found for the non-dimensional velocity of the overflow in these experiments. In Section 11.6 the difficulties in obtaining consistent results in such experiments with surface flows have already been noted, but these experiments were carried out in very large tanks, where unaccountable surface effects should be small.

Although the experimental results agree fairly well with the theoretical forecast for the overall advance of the front, in fact, the assumption of block flow makes no allowance for the maximum internal velocities. These have been found experimentally to be of the order of 1.2 times the front velocity; the relationship with the mixing at the leading edge has been investigated by later experimenters [4,5].

12.1.1 Trapezoidal, circular and other cross-sections
Some experiments carried out in trapezoidal flumes can be seen to fit quite well into the theoretical pattern.

In the case of the overflow front in a triangular section with a free surface, the initial velocity is related to the maximum depth by

$$U_0/(g'H)^{\frac{1}{2}} = 0.40$$

Between rectangular and triangular sections there is a reversal of the comparative pattern of underflow and overflow. Exchange flows in circular tubes are clearly also of importance and Fig. 12.2 gives values of the non-dimensional initial velocities in channels of rectangular, triangular and circular sections.

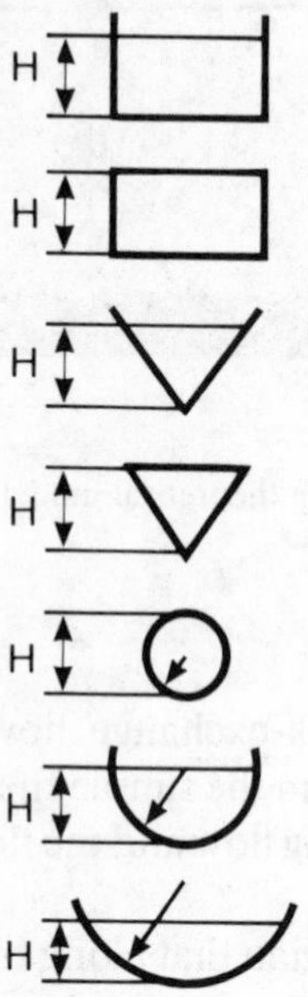

Section		Non-dimensional initial velocity	
		Underflow	Overflow
H	Open rectangular	0.465	0.59
H	Closed rectangular	0.44 approx	0.44 approx
H	Open triangular	0.67	0.41
H	Closed triangular	0.65	0.34
H	Circular pipe (full)	0.495	0.495
H	Half depth circular	0.51	0.44
H	Quarter depth circular	0.48	0.36

Fig. 12.2 — Experimental values of the initial velocity of gravity current fronts in channels of different cross-section. The velocity is given as $U/(g'H)^{\frac{1}{2}}$. (Courtesy of D.I.H.Barr).

12.2 RELEASE OF A FIXED QUANTITY OF FLUID IN A RECTANGULAR CHANNEL

The virtue of lock-exchange experiments is their relative simplicity, making it possible to perform large numbers of experimental runs to illustrate the whole range of gravity currents, including those in which viscous effects have become dominant. In many experiments the volume of one of the fluids is made much smaller than that of the other, so that the procedure approximates to the release of a finite volume of fluid into another fluid of infinite volume.

It has been shown experimentally that a gravity current produced by instantaneous release passes through two distinct phases. (A third phase may be reached if viscous effects become dominant.) After the rapid initial collapse when the gate is removed, there is an adjustment phase in which the front advances at constant speed. In this phase the initial conditions are important. This merges into an eventual phase in which the front speed decreases as $t^{-\frac{1}{3}}$ (where t is the time measured from release). It may be more convenient to express this relationship in the form "the distance, x, varies as $t^{\frac{2}{3}}$". The transition from the first to the second phase is observed to be rather abrupt.

12.2.1 First phase (constant speed)

It has been shown [6] that the transition from the first to the second phase occurs when a disturbance generated at the end wall (or plane of symmetry) overtakes the front.

The nature of the travelling disturbance differs in the following two cases: (a) those in which the segment of dense fluid has the same depth as the rest of the fluid, and (b) those where the dense fluid released is shallower than the total depth of the fluid, as in the "dam-break analogy" experiments.

In the former case the depth, h_0, of the dense fluid is equal to that of the rest of the fluid, H, in the channel. As soon as the gate is removed, the fluid from behind the gate forms a gravity current with structure as described above. The intense mixing between the two fluids is confined to a region just behind the leading edge of the current, the mixed fluid being left behind the head and above the following current. At the same time the displaced upper fluid forms a gravity current that propagates towards the end wall; see Fig. 12.3. When the backflowing current meets the wall, a hydraulic drop is generated [6]. This may be pictured as an "inverted bore" consisting of the lighter fluid advancing through the upper part of the dense fluid beneath it. This disturbance propagates away from the wall, and eventually overtakes the front. Some measurements of the position of the front of a volume of salt solution released from a lock in which $h_0/H = 1$ are shown in Fig. 12.4. In these experiments the front is observed to be about ten lock lengths from the end wall when it was overtaken by the bore. After this stage is reached the speed of the front is no longer constant and decreases with $t^{-\frac{1}{3}}$ (or x varies with $t^{\frac{2}{3}}$).

The behaviour of a "dambreak analogy" gravity current in deep water is different. Fig. 12.5 illustrates the difference from the lock-exchange in which a bore is formed; this is shown in Fig. 12.5(a). Fig. 12.5(b) shows the stages in the collapse of a volume of heavy fluid released into deep water, in which h_0/H approaches zero. As soon as the gate is removed, the dense fluid forms a gravity current moving away from the end wall at constant speed as in the previous case. At the same time a long wave of depression propagates along the fluid interface towards the end wall. This wave is reflected by the end wall (figure ii) and propagates away from the wall with speed slightly greater than the speed of the front (figure iii), eventually overtaking the front (figure iv). Thereafter, the front speed, which had been constant up to this point, decreases as $t^{-\frac{1}{3}}$ until viscous effects become more important than inertial effects, causing the front speed to decrease more rapidly.

The shallower the upper layer is relative to the dense layer, the greater must be the speed of the return flow there. Theory indicates that this trend results in the

(a)
(b)
(c)

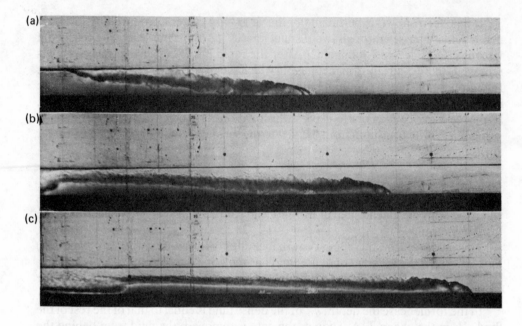

Fig. 12.3 — Shadowgraphs of a volume of salt water collapsing into fresh, at (a) 5, (b) 8 and (c) 11 seconds after release. The dotted vertical line shows the position of the lock gate. The vertical lines are 10 cm apart.

formation of an interfacial hydraulic drop for $h_0/H > 1/2$, but experiments give no indication of a hydraulic drop until $h_0/H > 0.7$. The values of the constant speed during the first stage for values of h_0/H other than 1 have been shown to decrease almost linearly from 0.7 when $h_0/H = 0$ to about 0.5, as already mentioned, for $h_0/H = 1$.

12.2.2 Second phase (self-similar flow)
During the stage when the front speed has begun to decrease, the gravity current is well described as collapsing through a series of equal-area rectangles, the so-called *box-model*, in which the current depth is roughly uniform along the length of the current, but steadily decreases with time.

On the grounds that the length of the current greatly exceeds its vertical thickness, an analysis has been based [7] on the depth-averaged shallow-water equations. Retaining only the buoyancy and inertial terms, the equations have been solved and the coefficients evaluated from various experiments, including some on the spreading of oil over water.

$$X = 1.6(g'q)^{\frac{1}{3}}t^{\frac{2}{3}}$$

was obtained for the two-dimensional case. This result has already been mentioned above as following the initial constant speed regime, and Fig. 12.6 gives some experimental results [10] showing the position of the front with time. In this log/log

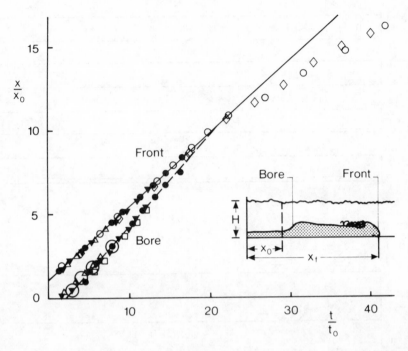

Fig. 12.4 — The position of a gravity current front which is being overtaken by a bore reflected from the end wall of a tank.

plot the distance $(x_t - x_0)$ is non-dimensionalised by the initial lock-length, x_0, and the time t by $t = (x_0/(g'h)^{\frac{1}{2}}$. It can be seen that the points during the initial adjustment phase lie on the line with slope 1 , showing uniform velocity. In the inviscid self-similar phase the points lie on the line with slope 2/3, showing distance varying with $t^{\frac{2}{3}}$.

The points representing flows that have begun to be dominated by viscosity can easily be distinguished when they reach this stage in the graph. They then slow down further and can be seen on lines with slope 1/5, a result to be discussed further in Chapter 15 in the context of viscosity-dominated currents.

12.2.3 Motion of air cavities in long horizontal ducts

Experiments in which a fixed volume of air is released into a closed horizontal tank full of water are illustrated in Fig. 12.7. Such experiments show clearly the stages in the development of a "gravity current" of air, after release from behind a gate, as it progresses above the water in a tank of 10 cm square cross-section.

In phase A the front is moving at a constant speed and occupies almost exactly half the depth of the tank. There is no turbulence at the head in this constant speed phase with no energy loss which was described theoretically in Chapter 11 and illustrated in Fig. 11.4.

In the second photograph, B, an advancing hydraulic jump, or bore, is seen on the left-hand side. This turbulent step in height of the water is steadily catching up the front.

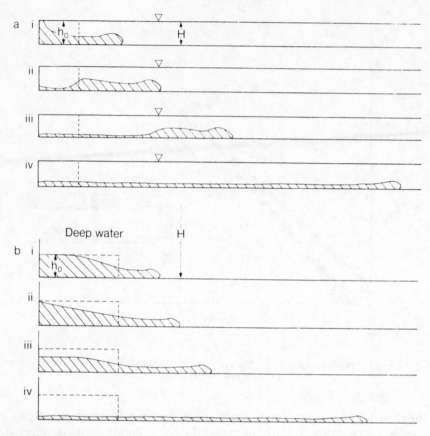

Fig. 12.5 — The collapse of a volume of dense fluid at four stages after release. (a) The volume of dense fluid has the same depth as the fresh water. (b) It is much less deep than the water in the channel.

In the third view, C, the bore has reached the head and a roughly level current of air exists, showing some turbulence at the lower interface with the water. The speed in this phase is no longer constant.

In four experiments (not shown here) the distance is was plotted against time. The distance, X, is non-dimensionalised by X_0, the lock-length, and the time by t_0, which is $(H/g')^{\frac{1}{2}}$. The results from the experiments all fell on the same lines. The bore is seen to be derived from the reflection from the end wall of the tank, and the end of the constant speed regime started when the bore had reached the front of the air flow.

There is also a third phase, which may be reached if the flow continues far enough. In this phase the cavity motion becomes erratic due to the dominance of surface tension forces [7]. The surface tension forces becomes strong enough to bring the motion to rest, i.e., the cavity approaches the limit of a static two-dimensional bubble. The motion becomes erratic because the randomly distributed contaminants on the plexiglass surface affect the contact angle.

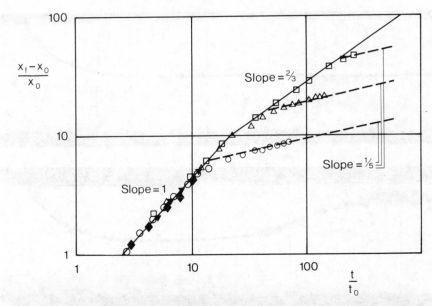

Fig. 12.6 — The position of four different gravity fronts with time. The change of gradient from 1 to 2/3 occurs when the distance is 10 lock-lengths. The graphs of the three currents which attain the viscous regime change at successive times to a gradient of 1/5.

12.3 AXISYMMETRICAL COLLAPSE OF A FIXED QUANTITY OF FLUID

A fixed quantity of dense fluid released into another fluid of different density, with no restraining barriers, spreads out in all directions. In early experiments [8] a cylinder of dense fluid was released into a larger tank. After a rapid initial collapse a roughly axisymmetrical spread of the fluid was observed.

After an initial adjustment phase, but before a viscosity-controlled stage is reached, a cylindrical "box model" is appropriate and it has been shown [9,10] that the distance, X, travelled by the front varies with the time, t, to the power of $\frac{1}{2}$.

Later experiments on axisymmetrical flows have been made using tanks as shown in Fig. 12.8 in which the flow in a sector of about 10° of a cylinder is observed. This uses much less volume of fluid for any given distance travelled, and has the advantage that in a transparent sector tank the profile of the head of the front can be observed more accurately than in a complete cylinder. It is also possible to take shadowgraph photographs.

The line of the front becomes nearly straight as a radial flow develops, and, as might be expected, the head then behaves very similarly to that observed in a flow between parallel walls. One difference is that the distance from the start is proportional to $t^{\frac{1}{2}}$ instead of $t^{\frac{2}{3}}$.

12.3.1 Initial vortex ring

In the initial phase of adjustment, however, there are important differences beteeen the two-dimensional and axisymmetric flows. Fig. 12.9(i) and (ii) shows sequential shadowgraphs of the initial collapse of salt water in a sector-shaped tank. In (i) the

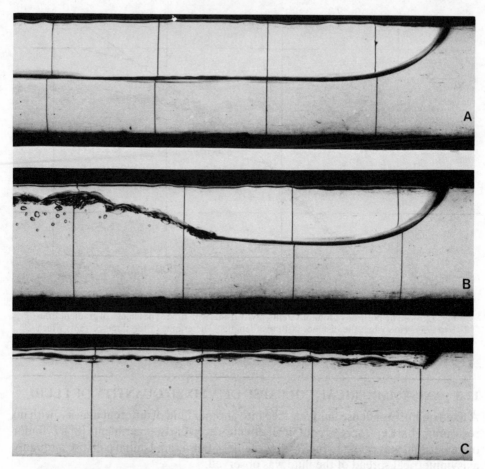

Fig. 12.7 — Three stages of the form of an advancing volume of air which has been released from one end of a channel of water. The depth of the tank is 10 cm.

initial fractional depth h_0/H is one quarter, and in (ii) $h_0=H$. In all these flows, and indeed also in the two-dimensional releases, the front forms very soon after release (in a few tenths of a second) and the most intense mixing of the two fluids occurs near the front. The most striking difference between the axisymmetric and two-dimensional flows is the intensity of rotational motion of the internal fluid during the early stages. In the time taken by the axisymmetric front to travel a distance equal to about one lock-length, x, the majority of the fluid in the current becomes concentrated at the front (for small h_0/H) or in multiple fronts (for h_0/H near to 1), leaving only a thin layer of heavy fluid near the ground. In contrast the two-dimensional flows are more uniform in depth shortly after release. In addition, the internal rotational flow and associated mixing are more intense; indeed, the mixing appears to occur all the way down to the ground behind the front (or fronts) in these flows.

The formation of this vortex ring during the early stages of spreading out of a dense fluid has important applications to aircraft safety and was dealt with in more

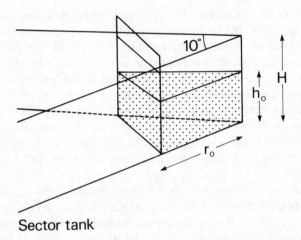

Fig. 12.8 — Tank in the form of a sector for the release of a fixed volume of dense fluid to form an axisymmetrical gravity current.

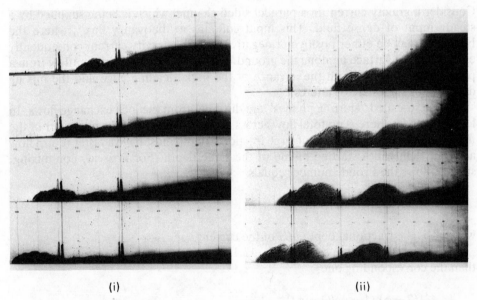

(i)   (ii)

Fig 12.9 — Shadowgraphs of a volume of salt water collapsing into fresh, in a sector-shaped tank. Total depth is 40 cm and reduced gravity, g', is 47 cm/s^{-2}. (i) $h_0/H=0.25$; (ii) $h_0/H=1$.

detail in Chapter 6. The details of this important vortex development are shown in more detail in Fig. 12.10. In these photographs the flow patterns are made visible by the suspension of fine aluminium particles in the dense fluid. The pictures show that by stage (c) the vortex has extended down close to the lower boundary.

We know that Kelvin–Helmholtz vortices form as the leading edge of a gravity current develops, and because the fluid volume in a vortex is approximately conserved, its cross-sectional area must decrease as it stretches. Conservation of angular momentum about the centre line of the vortex then implies that its intensity increases. This intensification is largest during the rapid expansion near the source and produces a leading edge vortex which occupies almost the full depth of the dense fluid.

To show that the vorticity already being produced at the leading edge of an advancing gravity current is sufficient to produce this intense roll-up of the leading edge when it is stretched, a simple experiment has been carried out [11]. In this experiment a saline gravity current (produced by lock-exchange) flowing between parallel walls was suddenly allowed to double its width. Fig. 12.11 shows the simple apparatus. The behaviour of the leading edge of the gravity current is shown in Fig. 12.12 as it was allowed to double its width. The width of the tank increased from 10 cm to 20 cm between the two white lines which are 30 cm apart. The leading edge vortex rolled up as the width of the front increased and reached almost down to the ground in Fig. 12.12(b).

12.4 CONSTANT FLUX IN PARALLEL CHANNEL

Consider a gravity current in a parallel-sided channel which is being supplied by a steady input of dense fluid. This input contains a "buoyancy flux", where the buoyancy may be either positive or negative, and the gravity current consequently either at a free surface or along the ground. An example is the warm outflow from a power station where both the advance of the initial leading edge and the mixing during later stages are of interest.

These so-called "starting plumes" are different from the lock exchange flows. In lock exchanges there is no total flow across any cross-section of the container of the fluid, so that a gravity current of large fractional depth must have a faster return flow above it than will a starting plume of the same depth. For inviscid, non-mixing, starting flows the Froude number equals

$$\{(2 - \phi)/(1 - \phi^2)\}^{\frac{1}{2}}$$

where ϕ is the fractional depth h/H of the current.

As shown in the upper curve of Fig. 12.13 this expression varies much less with ϕ than the corresponding one

$$\{(2 - \phi)(1 - \phi)/(1 + \phi)\}^{\frac{1}{2}}$$

for lock-exchange flows, shown in the lower curve.

It is possible to set up an experimental flow rate in which the layer stays uniform for some distance along the plume but for increasing distances the velocity decreases.

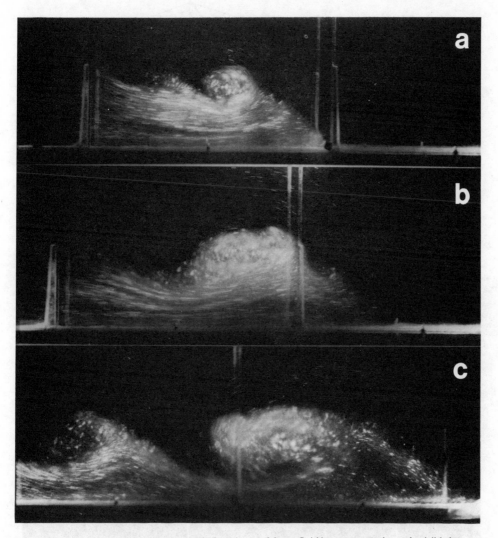

Fig. 12.10—The collapse and spread of a volume of dense fluid in a sector-tank, made visible by fine aluminium particles. Time intervals are 2 s and $g' = 12$ cm s^{-2}.

It has been shown that for an inlet Froude number greater than 1 there exists an entraining hydraulic jump at the interface between the fluids, as shown in Fig. 12.14 and it seems that the rate of entrainment here and the conditions downstream are controlled by the gravity-current head.

This hydraulic jump, or "density jump" as it has been called [12], is able to entrain the varying amounts of ambient fluid necessary to satisfy a range of downstream conditions. However, along a horizontal floor with a gravity current head no equilibrium state exists [13] and the velocity of the front continues to fall, eventually causing the density jump to flood.

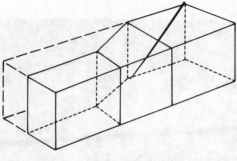

Widening tank

Fig. 12.11 — Tank to test the effect of widening on the form of the head of a gravity current, originally in a channel with parallel sides.

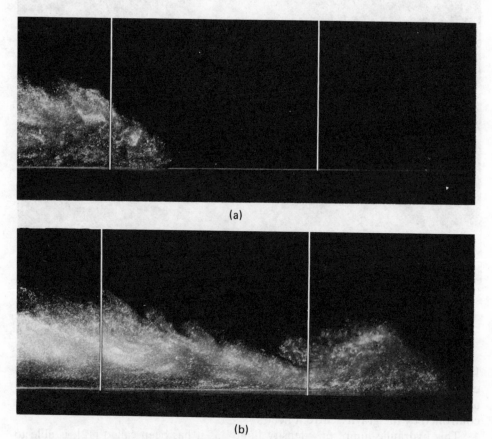

(a)

(b)

Fig. 12.12 — The effect of widening the channel on the form of a gravity current head. The width increases from 10 cm to 20 cm between the two white lines which are 30 cm apart. The time interval is 5.7 s, and g' is 14 cm s^{-2}.

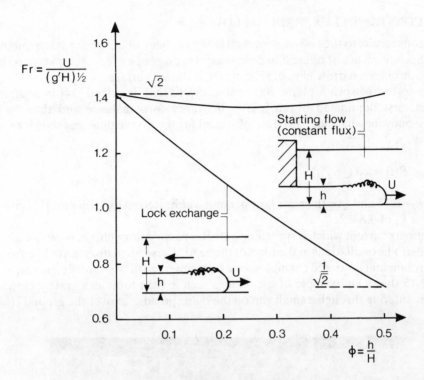

Fig. 12.13 — Comparison between the speed of the front of a lock-exchange flow and that of a constant flux, at the same fractional depth.

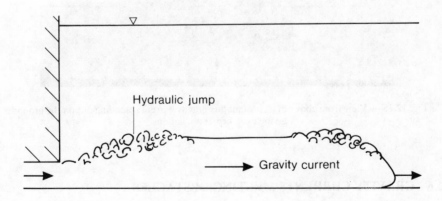

Fig. 12.14 — Gravity flow with inlet Froude number greater than one, showing the entraining hydraulic jump.

12.5 CONSTANT FLUX, RADIAL FLOW

Analogous results to the two-dimensional flow have been obtained for axisymmetric flows. Such flows are of interest in describing several geophysical events such as the spread, through a narrow gap, of river water at the sea surface.

The spread of such a plume has been analysed and described [14] in separate regimes, first dominated by an inertia–buoyancy force balance and then by a friction–buoyancy balance. The rate of spread for the first regime was shown to be given by

$$R(t) = k_1(Qg')^{\frac{1}{4}} t^{\frac{1}{2}}$$

where $k_1 = 0.75$ and Q is the mass flow per unit width. Recent experiments [15] give a value for k_1 of 0.84.

A gravity current which is spreading out after passing through a narrow slot often shows periodic oscillations in the form of the head. These oscillations have been seen in environmental flows and examples have been observed in laboratory experiments. Fig. 12.15 shows an example of oscillations seen in plan form as a gravity current which is entering through a small gap on the right spreads out over the ground [16].

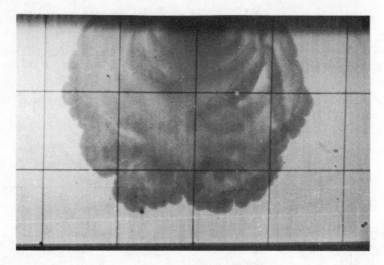

Fig. 12.15 — View from above of a constant flux gravity current spreading out on the ground, giving evidence of oscillations.

12.6 GRAVITY CURRENTS MEETING OBSTACLES

The gravity currents considered so far flow along a plane surface, either horizontally or at a uniform slope. The problem of the effects of barriers on the behaviour of gravity currents has wide practical applications, for example on the control of escaping dense gases or liquids. This section will consider some laboratory experiments which examine the effects of various kinds of obstacles.

12.6.1 Gravity current meeting a solid barrier.

A sketch of the flow under consideration is shown in a Fig. 12.16. A simple solution of this can be obtained, assuming that h_1 is constant during the interaction. The result predicts a hydraulic jump propagating away from the wall after the interaction.

This is not adequate to explain all the experimental results which show an initial jet of fluid that shoots up the wall on impact. The resulting "splash" is shown in Fig. 12.17 in four sequential photographs from an experiment using saline flows in a water tank [17]. In this experiment, with an "infinitely high" barrier, the high splash of heavy fluid runs up the wall to about twice the original height of the released fluid (this height is indicated by a dark horizontal line in the photographs). The last photograph shows a hydraulic jump moving away from the barrier.

Experiments with a barrier of height less than twice the height of the current were also performed. These all showed a similar splash, a reflected hydraulic jump, but some of the fluid continued to flow on beyond the barrier and formed a gravity current on the other side.

The nature of the return flow resulting from the reflection is actually more complex than in the simple reflection from a wall. Firstly the raised head of the gravity current gives an initial reflected surge which is deeper than the following steady reflected flow. Secondly the mixing which has been taking place at the head during the advance of the current leaves a stable stratified layer above the dense current, as described in Section 11.4. These two properties result in the hydraulic jump being manifested as primarily a smooth solitary wave moving back through the stable layer, progressing at a speed close to that of the original gravity current, but in the opposite direction.

Records made of the progress of the reflections of gravity currents from the end walls of laboratory tanks show that the first reflected disturbance travels back at a uniform speed not very different from that of the original gravity current. Fig. 12.18 shows a disturbance produced by reflecting a gravity current from the end wall of a tank. The limited amount of mass, seen moving to the left, is small and it much resembles a "solitary wave". The presence of the dye patch, which was inserted at the end wall, makes it clear that some of the original mass is still being carried forward.

Experiments have also been done with barriers of other forms. For example, barriers of dykes of different slopes have been employed and the dependence of spillage on the inclination of the dyke has been determined experimentally [18].

12.6.2 Gravity current flowing through a porous obstacle

The series of photographs in Fig. 12.19 shows the interaction of a gravity current with a porous barrier consisting of 40 wooden dowels, evenly spaced. Firstly the gravity current increases in depth as it encounters the barrier, and it can be seen that its height nearly doubles. The speed of the front is also retarded. Soon, however, the heavy fluid seeps through at the bottom of the barrier and begins to form a gravity current front, while at the same time a weak hydraulic jump propagates upstream from the barrier. The splash associated with the initial encounter is much weaker, as expected, than for the case with the solid barrier. Eventually, a steady state is achieved with a rapid drop in the interface level through the barrier with a gravity current front propagating downstream away from the barrier.

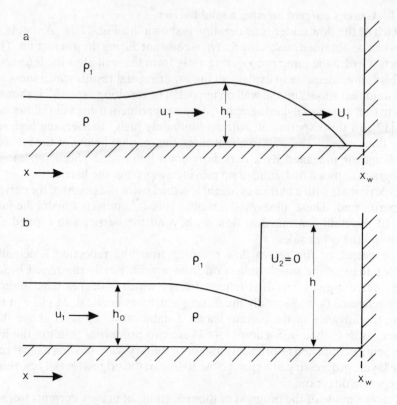

Fig. 12.16 — Schematic illustration of the interaction of a gravity current with a solid vertical wall.

12.6.3 Flow under a barrier

Fig. 12.20 shows the form of the head of a laboratory gravity current which has just passed beneath a sharp barrier across the upper half of a two-dimensional channel. This current was generated by a lock-exchange and as it approached the barrier the dense fluid occupied about half the depth of the tank. The shadowgraph shows that the depth of the current was reduced to about half the height of the gap beneath the barrier.

One feature which is clearly shown in this shadowgraph is the turbulence formed above the dense current as it approaches the barrier. The reverse flow through the restricted space breaks up into a turbulent zone filling most of the space up to the surface.

A rather similar environmental problem which can be examined in the laboratory is that of a power station effluent in which a gravity current of warm water on the surface of the sea passes above a submarine ridge. To avoid problems of a contaminated free surface and uncontrolled heat losses, the laboratory experiment shown in Fig. 12.21 has been carried out, in which the flow and obstacle have been inverted. A dense salt solution moved beneath a smooth obstacle, shaped as shown,

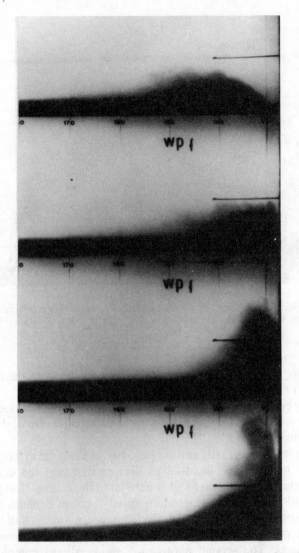

Fig. 12.17 — Four views of an experimental collision of a gravity current front with a vertical wall, showing the "splash" which moves up the wall. The original height of the dense fluid before release is shown by the black horizontal line.

and a series of photographs served to record the progress of the front and to show its changes in form.

As the gravity current passed beneath the obstacle it became reduced to half its original height and measurements showed that its speed was reduced by the factor of $1/\sqrt{2}$.

12.6.4 Gravity current in steadily decreasing depth

Experiments on gravity currents flowing into fluid of steadily decreasing depth are best carried out with dense currents running along the floor, beneath a rigid sloping lid. As in the previous section, surface contamination effects can be avoided and it is

Fig. 12.18 — Shadowgraph of the disturbance produced by reflecting a gravity current from the end wall of a tank. The appearance is close to that of a solitary wave.

possible to apply a small correction to the results for flows in the ocean at a free surface, above a rising sea bed.

Each gravity current was released from a lock at the end of the tank and allowed to flow beneath a rigid flat ceiling sloping down in the direction of the current. The angle of slope varied from 1° to 21°.

When the depth of the gravity current head became as large as half the local depth below the lid, the breaking head-waves disappeared and the current interface downstream of the head became roughly parallel to the ceiling. The photographs in Fig. 12.22(a) and (b) show two stages of a gravity current beneath a lid with slope about 5°. The depth of the dense fluid at the front remains about half the local depth, so that the depth of the parallel-sided space above the liquid continues to decrease as the front advances. The second photograph shows almost complete absence of mixing at the interface.

Measurements were made of the speed, U, of the front for different values of the horizontal distance, x, from the intersection of the ceiling and the floor, for different ceiling angles. Fig. 12.23 shows values of U^2 plotted against $g'x \tan \theta$ and it can be seen that the points lie close to a straight line through the origin, giving $U^2 = 0.21 g'x \tan \theta$, or $U = 0.45 (g'x \tan \theta)^{\frac{1}{2}}$.

As the form of the head under the sloping lid resembles the non-mixing head seen at low Reynolds number gravity currents, tests were made to see whether the non-mixing head was a low Reynolds number effect. In one such test the sloping lid was made parallel to the floor before the end of the tank; the usual head-structure with breaking waves appeared beneath the level lid. The least slope which would suppress head-mixing was found to be about 1/80, and all the experimental values of the Froude number $U/(g'x \tan \theta)$, for slope greater than 1/80, were plotted against the

Fig. 12.19 — The interaction of a gravity current with a porous obstacle. This consists of 4 rows
of 10 rods, 10 mm in diameter and 10 cm high.

Reynolds number Re $= Ux \tan\theta/v$. No trend with Reynolds number could be
detected in the range from Re = 50 to 5000 examined.

If we write $H = x \tan\theta$, the total height of the space immediately above the front
of the current, then it appears that $U = 0.45 (g'H)^{\frac{1}{2}}$ and the current at any time adjusts
almost instantaneously to the same value as it would have underneath a level lid of
the same height, since $U = 0.45 (g'H)^{\frac{1}{2}}$ is the usually accepted value for the speed in
the initial constant speed regime in a lock-exchange flow of total depth H.

BIBLIOGRAPHY

[1] O'Brien, M. P. & Cherno, J. 1934. Model law for motion of salt water through
    fresh. *Trans. Am. Soc. Civil Eng.*, 99: 576–594.
[2] Yih, C-S. 1980. *Stratified Flows*. Academic Press, New York pp 204–207.

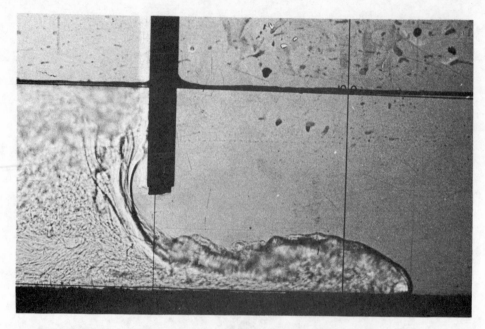

Fig. 12.20 — The form of the head of a gravity current which has just passed beneath a sharp barrier.

[3] Barr, D. I. H. 1967. Densimetric exchange flow in rectangular channels. III. Large scale experiments. *Houille Blanche*, 6/1967: 619–631.

[4] Barr, D. I. H. & Hassan, A. M. H. 1963. Density exchange and flow in rectangular channels.II. Some observations of the structure of lock exchange flow. *Houille Blanche*, 7/1963: 757–766.

[5] Simpson, J.E. & Britter, R.E. 1979 The dynamics of the head of a gravity current advancing over a horizontal surface. *J. Fluid Mech.* 94: 477–495.

[6] Rottman, J. W. & Simpson, J. E. 1983. Gravity currents produced by instantaneous releases of a heavy fluid in a rectangular channel. *J. Fluid Mech.* 135: 95–110.

[7] Baines, W. D., Rottman, J. W. & Simpson, J. E. 1985. The motion of constant volume air cavities in long horizontal tubes. *J. Fluid Mech.*, 161: 313–327.

[8] Penny, W. G. & Thornhill, C. K. 1952. The dispersion under gravity of a column of fluid supported on a rigid horizontal plane. *Proc. Roy. Soc. London*, A244: 283–311.

[9] Hoult, D. P. 1972. Oil spreading on the sea. *Ann. Rev. Fluid Mech.*, 4: 341–368.

[10] Fay, J. 1969. The spread of oil slicks on a calm sea. In: *Oil on the Sea* (ed. D. P. Hoult), Plenum Press, pp. 43–63.

[11] Rottman, J. W. & Simpson, J. E. 1983. The initial development of gravity currents from fixed-volume releases of heavy fluids. *IUTAM Symposium*, Delft, 1983, pp. 347-359.

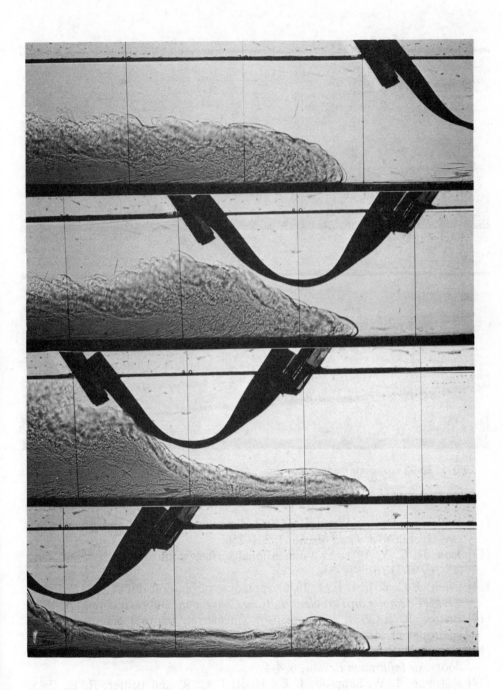

Fig. 12.21 — Stages in the passage of a gravity current beneath a smooth barrier occupying half the depth of the tank. The lines are 10 cm apart, and the times since release are 8.4, 11.5, 14.4 and 17 seconds.

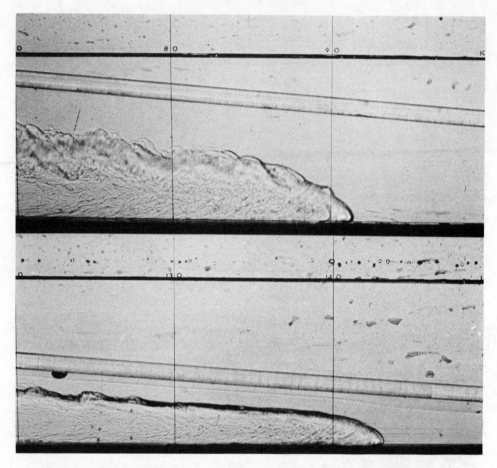

Fig. 12.22—Two stages in the flow of a gravity current beneath a sloping lid. The lines are 10 cm apart, and the times since release are 17.4 and 25 seconds.

[12] Wilkinson, D. L. & Wood, I. R. 1972. A rapidly varying flow phenomenon in a two-layer flow. *J. Fluid Mech.* 47: 241–256.

[13] Koh, R. C. Y. 1971. Two-dimensional surface warm jets. *J. Hydraul. Div. ASCE,* 97(HY6): 819–836.

[14] Chen, J.-C, & List, E. J. 1976. Spreading of buoyant discharges. *Proc. 1st CHMIT Seminar on Turbulent Buoyant Convection,* Dubrovnik, pp. 171–182.

[15] Britter, R. E. 1979. The spread of a negatively buoyant plume in a calm environment. *Atmos. Environ.* 13: 1241–1247.

[16] Linden, P. F. & Simpson, J. E. 1985. Buoyancy driven flow through an open door. *Air Infiltration Review,* 6: 4–5.

[17] Rottman, J. W, Simpson, J. E., Hunt, J. C. R. and Britter, R. E. 1985. Unsteady gravity currents over obstacles. *J. Hazardous Materials,* 11: 325–340.

[18] Greenspan, H. P. & Young, R. E. 1978. Flow over a containment dyke. *J. Fluid Mech.,* 87: 179–192.

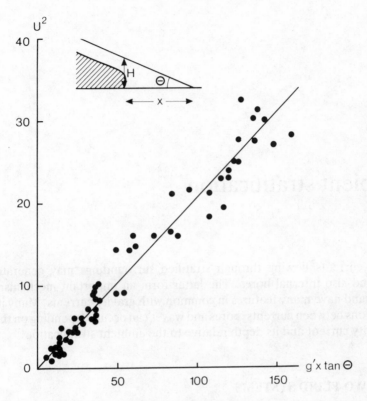

Fig. 12.23 — The square of the speed plotted against distance for gravity currents travelling beneath a sloping lid.

13

Ambient stratification

Gravity currents flowing through stratified surroundings may generate internal waves and also internal bores. The latter form an important mechanism of mass transfer and have many features in common with gravity currents. Many interesting interactions between currents, bores and waves can occur, depending on the speed of the gravity current and its depth relative to the ambient stratification.

13.1 TWO-FLUID SYSTEMS

A two-layer system will be considered first in which one fluid of uniform density lies above another of appreciably greater density, with a sharp interface between them. The behaviour of a gravity current which moves along the lower boundary beneath these fluids depends on the nature of the disturbance that its progress creates at the interface between them.

There are several possibilities, and it will be simpler to consider first the effects of moving a solid obstacle of similar shape at different speeds through the stratified system. It will then be possible to see how a gravity current in such a system can sometimes generate an internal bore moving ahead of it.

13.1.1 Internal bore formation

The types of disturbance which may be generated at the interface by a moving obstacle are displayed in Fig. 13.1. This shows four different flow patterns which can result from a start-up after a state of rest. They are defined in terms of the two non-dimensional quantities $F_0 = U/(g'h_0)^{\frac{1}{2}}$ and $H = d_0/h_0$, representing the *speed* and *height* of the obstacle, h_0 and d_0 being the heights of the fluid interface and the obstacle, respectively. The results apply strictly to an upper fluid of infinite depth [1].

On the left of the inset graph are found flows (a) and (d) involving obstacles whose depth is small compared with the depth of the lower fluid . The effect produced by these smaller obstacles consists only of a simple displacement of the interface either upwards or downwards, according to whether the flow is *supercritical* or *subcritical*. That is according to whether its velocity, U, is greater or less than

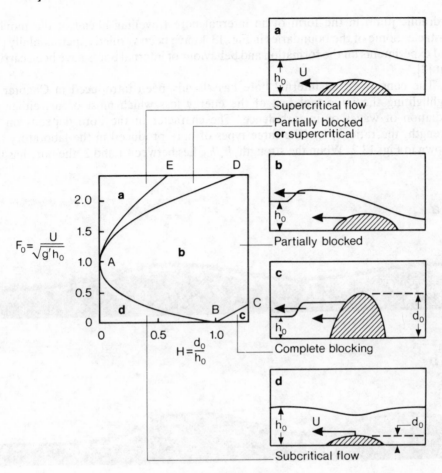

Fig. 13.1 — Four types of disturbance which can be generated by a moving obstacle at the interface between two fluids.

$(g'h_0)^{\frac{1}{2}}$, which represents the maximum speed of small internal waves on the interface. Another way of stating this is "whether the internal Froude number $U/(g'h_0)^{\frac{1}{2}}$ is greater or less than 1".

When the height of the obstacle approaches the depth of the layer, as in (b), or exceeds it, as in (c), an upstream influence in the form of a hydraulic jump appears for an increasing range of speeds. Partial *blocking* may exist, in which the flow above the obstacle is reduced and some of the fluid is forced to move ahead of the obstacle. Blocking of a flow past a stationary obstacle is usually defined as the stagnation of a layer of the fluid leading upstream from the obstacle. So, in the case of a moving obstacle, blocking is taken to mean the movement of a region of fluid upstream at the speed of the obstacle. This is not synonymous with the presence or absence of the upstream influence of the obstacle. In a stratified flow it is possible for the part of the fluid near the bottom to be blocked, while the top part still flows.

In all these cases, the partial or complete blocking results in the generation of a hydraulic jump in the form of an internal bore travelling ahead of the moving obstacle. Some of the boundaries in Fig. 13.1 have been verified experimentally [2], and experiments on the formation and behaviour of internal bores have been carried out [3].

The concept of an internal bore has already been introduced in Chapter 1, highlighting the important fact of the energy loss which must occur, either by radiation of waves or in turbulence. The character of the bore depends on its strength, the ratio h_1/h_0. The three types of bore produced in the laboratory are shown in Fig. 13.2. When the strength, h_1/h_0, lies between 1 and 2, the bore has the

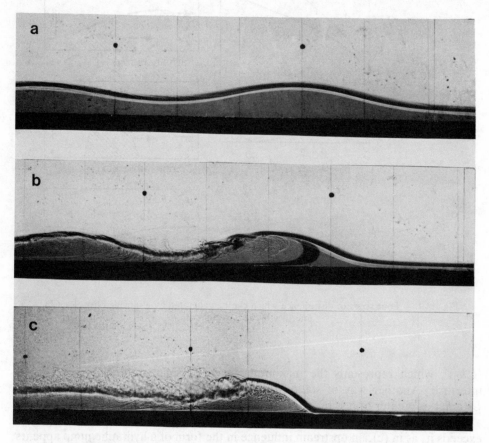

Fig. 13.2 — Three different forms of internal bore seen at an interface between two fluids. (a) Strength between 1 and 2. (b) Strength between 2 and 4. (c) Strength greater than 4.

smooth undular form shown in the first photograph (a). The undulations are produced over the obstacle one after another and travel upstream at the bore speed; little or no mixing occurs between the two fluids. When h_1/h_0 lies between 2 and 4, the bore has the form shown in the second photograph (b). In this case the bore is almost undular, but some mixing occurs on the downstream face of the first undulation, and also at points further downstream, due to shear instability. As h_1/h_0 gets close to 4, this mixing becomes more significant and can be seen on the downstream faces of the first few undulations. When h_1/h_0 is greater than 4 the mixing completely dominates the motion, obliterating any undulations, as shown in the third photograph (c) in Fig. 13.2. The bore then appears like a gravity current; this is easily seen by comparing this photograph with several different ones of gravity currents shown in Chapter 11.

Theories for bores in two-layer fluids have been developed when there is no mixing between the two fluids [3, 4, 5]. The case of greatest interest is that in which the upper layer is very deep compared with the lower layer; the result is then

$$F_0^2 = \frac{1}{2} \frac{h_1}{h_0} \left(1 + \frac{h_1}{h_0}\right)$$

This expression is plotted in Fig. 13.3 together with experimental results [3]. The

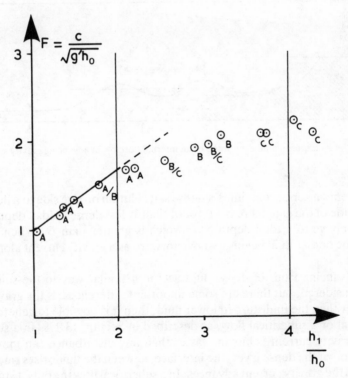

Fig. 13.3 — Experimental values of the variation of bore speed, c, with strength. The speed is non-dimensionalised in the form of a Froude number.

agreement is good to about $h_1/h_0 = 2$, but for larger values of this ratio the mixing between the two fluids reduces the speed below the theoretical value. When h_1/h_0 is greater than 2, the bore speed is quite well predicted by gravity current theory.

Attempts have been made to measure the wavelength in undular bores. Through the range of bore strengths from 1.5 to 2.5 the ratio of the wavelength, λ, to the depth h_1 of the bore is found to be approximately 10.

13.1.2   Generation of bore by gravity current

If a gravity current moves along the ground disturbing a dense layer of fluid, the flow relative to the head can be either supercritical or subcritical. It is more likely, however, that the current will give rise to a hydraulic jump in the form of an internal bore.

The nature of the disturbance produced depends on the relative densities and depths of the current and the layer. The generation of internal bores by gravity currents has been investigated in the laboratory [6], using the arrangement shown in Fig. 13.4. In this arrangement a gate is introduced into a water tank at a given

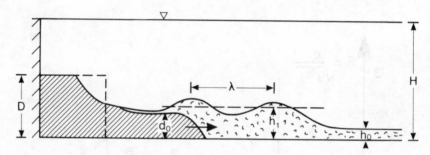

Fig. 13.4 — Experimental generation of an internal bore by an advancing gravity current.

distance from one end, containing a dense salt solution on one side to a depth D. On the other side of the gate there is a layer of slightly less dense fluid of depth h_0. Fresh water lies above to a total depth of H which is greater than D. When the gate is removed the denser fluid begins to move forward as a gravity current along the floor of the tank.

The advancing front of dense fluid acts in a similar way to the solid obstacle already considered, but there are some important differences. If the gravity current is shallower than the undisturbed lower fluid, then it is possible to generate either a supercritical or a subcritical flow, as described in Section 13.1.1. In a supercritical flow the gravity current is moving faster than any disturbance can move forward along the top of the dense layer; the interface between the fluids rises smoothly over the head as the gravity current advances. In a subcritical flow the only disturbance is a small depression in the layer which moves along above the advancing gravity current head.

When the depth, d_0, of the gravity current is greater than the depth, h_0, of the layer, the flow of the dense layer is partially blocked and a bore is generated, which may separate from the gravity current and move ahead of it. The first stage in the generation of an undular bore is the formation of a smooth hump which envelops the head of the gravity current. This hump moves forward along the interface between the two layers, leaving behind it some of the denser fluid of the gravity current. The gravity current is disrupted, but a fresh head soon forms, a fresh bulge forms above this, and the process is repeated.

The photographs in Fig. 13.5 show three stages in the development of an undular

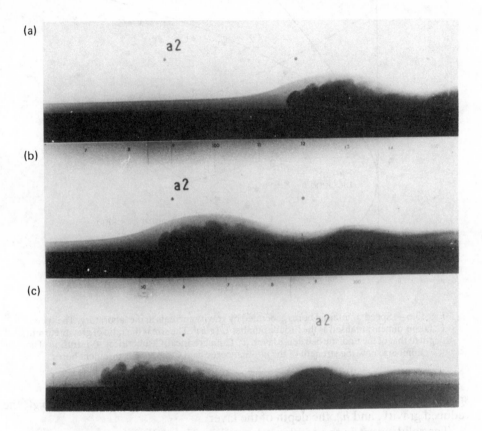

Fig. 13.5 — Three stages in the generation of an undular bore by an advancing gravity current.

bore by an advancing gravity current. In (a) the bore is starting to grow as a smooth hump enveloping the head of the gravity current. In (b) a second front and wave are beginning to take shape. In the third view, (c), some of the dense fluid beneath the first wave of the bore is starting to be left behind and a third bulge is starting to develop. Soon all this dense fluid will be left behind and the first wave of the bore will be completely separated from the gravity current. The process will continue as each successive wave separates from the gravity current.

Some experimental results [6] in Fig. 13.6 show the speed of experimental bores generated by gravity currents of various sizes. The sizes of the gravity currents which were forcing the bores are given in terms of the ratio of the depth of the gravity current, d_0, (which is approximately $\frac{1}{2}D$), to that of the undisturbed layer, h_0. The

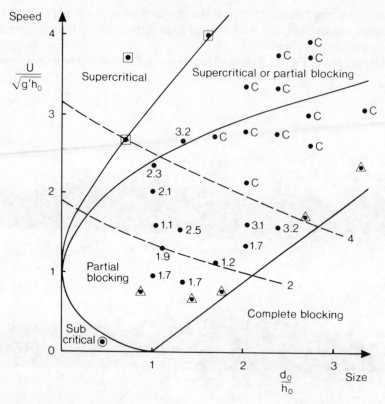

Fig. 13.6 — Speed of internal bores generated by gravity currents in the laboratory. The speed, U, is non-dimensionalised as the Froude number $U/(g'h)^{\frac{1}{2}}$. The size is the ratio of gravity current depth to that of the undisturbed dense layer, h_0. □ supercritical, ○ subcritical, △ intrusion. The numbers show the strength of the internal bores, and C represents turbulent bores.

speeds, U, are non-dimensionalised as a Froude number, $U/(g'h_0)^{\frac{1}{2}}$, based on g', the reduced gravity, and h_0, the depth of the layer.

The solid curves in the graph mark out the areas of disturbances expected to be formed by the movement of a solid obstacle, as shown in Fig. 13.1. Additional dashed curves show the expected boundaries of bores of strength 2 and 4. Super- and sub-critical flows are plotted as square and circular symbols. As expected, they appear with relatively small gravity currents in regions where the Froude number is greater and less than 1 respectively.

The points with numbers from 1.1 to 3.2 refer to the strengths of bores of undular form, as in the classes A and B in Fig. 13.2. Points marked with letters C refer to turbulent bores, with strength greater than 4 as shown in the same figure. If the gravity current has a depth greater than about four times that of the lower layer, this layer is mixed into the gravity current which then behaves almost as if the lower layer does not exist.

The area corresponding to "complete blocking" is blank, as it does not seem to be possible to achieve this state by a gravity current moving through a layer. The nearest flow situation that can be attained yields bores generated by intrusive gravity currents. These currents do not move along the floor, but travel along the interface between the two layers. Points corresponding to bores generated by intrusive gravity currents are marked with triangles and it can be seen that they lie close to the boundary of the "complete blocking" zone. Bores formed by intrusive gravity currents will be discussed below.

13.1.3 Intrusive gravity currents

Both in the ocean and in the atmosphere many examples of density layering exist which are thought to be caused by intermediate or **intrusive** gravity currents. These intrusions may occur after a limited section of two-layer fluid has been thoroughly mixed and released. A gravity current of mixed fluid flows along the interface, and furthermore this current can sometimes generate a bore moving ahead of it.

Intrusions along a sharp interface
A limited class of intrusive gravity currents can be produced very simply in the laboratory. The lower half of a tank is filled with salt solution, and tap water is carefully added to an equal depth above it. A partition is inserted near one end and the two fluids behind it are then stirred until they are completely mixed. When the partition is removed, the mixed fluid collapses and flows as an intrusion along the interface between the two fluids. In a refinement of this apparatus the intrusion is derived from a layer of fluid of intermediate density introduced at an appropriate time during the filling process.

The thickness of the interface between the two fluids, which always increases slowly with time due to molecular diffusion, has significant effects on the nature of the head of the intrusive current. The thickness of the interface between the two fluids in the intrusion shown in Fig. 13.7 was found to be just less than 0.2 of the thickness of the gravity current. It has been found experimentally [7] that this relative thickness can be taken as effectively "sharp". The speed of such a symmetrical intrusion varies with fractional depth in much the same way as an inviscid boundary gravity current.

Intrusive gravity currentswhich occur in nature may not always have exactly the mean density of the two layers. Asymmetric flows have features of more interest than the symmetrical ones, and these currents are capable of generating bores by displacing the level of a sharp interface. A range of internal bores can be produced in the laboratory by replacing the fluid behind the barrier with one of density either greater or less than the mean of the two layers and by altering the relative depth of the layers. Fig. 13.8 shows such a bore produced at a thin sharp layer by an intrusive gravity currrent with density only slightly greater than that of the lower fluid. In this picture the first undulation of the bore has moved about 10 cm ahead of the intrusion, and the second undulation has almost separated from the head of the intrusion.

Intrusions along a thick interface between two fluids
When the interface between the two fluids is "thick", the behaviour of an intrusive gravity current is best understood by considering once more the symmetrical case.

In small-scale laboratory experiments it may not be easy to maintain the thickness

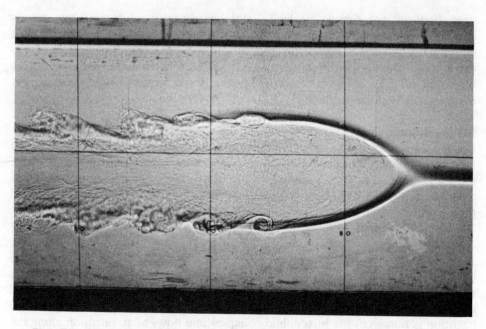

Fig. 13.7 — Shadowgraph of an intrusive gravity current, running along a sharp interface
between two fluids.

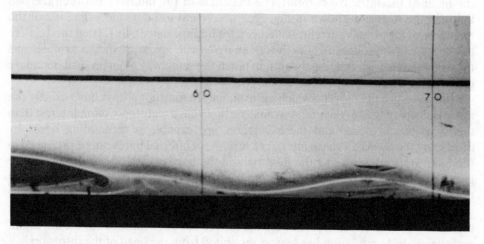

Fig. 13.8 — Experimental production of a bore by an intrusive gravity current.

of the interface at less than 0.2 of the intrusion, as molecular diffusion will
continually increase the width of the interface to a thickness which may be
significant.

If the interface thickness increases, the speed of advance of a symmetrical intrusion is increased. Fig. 13.9 shows some results [7] of experiments to measure the effect of the interface thickness on the speed of a symmetrical intrusion. The current through a thin interface shown in Fig. 13.7 is represented by the point A in the graph. As the relative thickness of the interface increases, the structure of the gravity current alters, and when the interface thickness is 0.7 of the gravity current (marked as point B) the head of the current has disappeared and the appearance is that of the reflection of a bore whose strength is 4 or greater, the type of bore which can be formed by a gravity current running into a thin dense layer.

The behaviour of such gravity currents running into an interface of appreciable thickness can be interpreted by replacing the curved interfacial density profile by a step profile. It can be proved that this is a legitimate substitution as regards the formation of bores provided that the thickness of the layer is less than the depth of the obstacle which is forcing the bore.

Using this approximation, all symmetrical intrusive gravity currents and bores can be interpreted in terms of flows with a dense layer above the floor, which have been completely reflected about the line of the ground.

Two examples are shown in Fig. 13.10 which should help to make this clear. Fig. 13.10(a) shows the generation of a turbulent bore, with the disappearance of a raised head. It should be possible to produce a "reflected undular bore" from gravity currents at lower speeds, as sketched in Fig. 13.10(b), and such bores have been observed experimentally, not only in parallel channels, but also in diverging radial flows.

The early stages of a bore of this type are shown in the photograph in Fig. 13.11, in which the bulge forming around the gravity current head almost completely separates the head of the dense fluid from the following flow.

Caution is needed in interpretation of such mixed layer experiments in which both the lock length and the distance travelled by the current are of limited length. If the lock length is short, then it is possible for the currents above and below, flowing in the opposite direction to the gravity current, to be reflected from the end wall and to catch up the front. If this happens, then the bore may be reduced to merely the appearance of a solitary wave. Also, in a short run of a bore, at the start of its generation, the appearance is that of a simple solitary wave.

Allied experiments [8] made in axisymmetric, or radial, flows show similar effects. Fig. 13.12 shows two stages in the development of a mixed region advancing through a two-level stratification, in which the development of a series of waves can be seen.

13.1.4   Collision of two gravity currents

If two gravity currents meet they may interact in various ways. One important example in the atmosphere is the collision between two mesoscale frontal flows. One important result which has been observed is the combination of the two areas of rising air to produce an especially strong zone of convection; another may be the production of a third unexpectedly strong gravity current.

There is also evidence from sedimentary deposits on the ocean bed of the collisions of two turbidity currents.

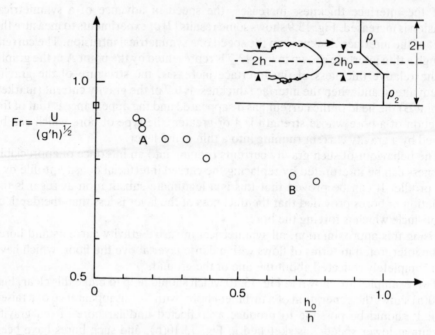

Fig. 13.9 — Experimental results which show dependence of speed on intrusions of the thickness of the interface between the two fluids.

The simplest case to study in the laboratory is the head-on collision between two saline gravity currents, flowing in a parallel-sided water channel. The results of such a collision between two gravity currents of different size and density are shown in the four photographs of Fig. 13.13 (a)–(d), in which the gravity current moving from the left is marked with a dark dye.

The form of each front soon after the collision, shown in (b), much resembles the shape of a gravity current being reflected from a vertical wall. This type of event was described in Section 12.6.1 and it was shown that the reflected disturbance was a hydraulic jump in the form of an undular bore. Developing bores appear in stage (c) , a large dark one on the left and a small light one on the right, moving away from each other. It is clear that the two gravity currents were not equal in height and density; the current from the left is reflected as a deep turbulent bore, but that from the right reappears as a weak one, of undular form, possibly only consisting of a solitary wave.

If the fluid in each current in the experiments were not distinguished from the other by the use of a dye, it would be easy to make the mistake that the two gravity currents had met and "crossed", and that each one continued to travel in the same direction as before.

"Crossing fronts" have been described in the atmosphere in which clear disturbances emerged from the meeting, and moved at about the same speed and direction as before the collision. It is likely, however, in the atmospheric encounters that only a small amount of the fluids actually crosses and that these events involve the kind of undular bore formation we have described.

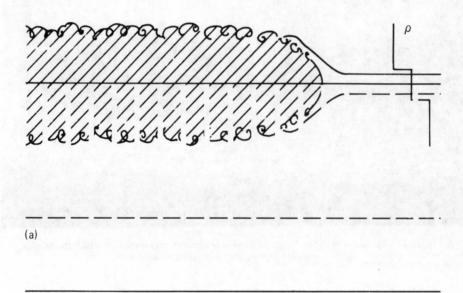

(a)

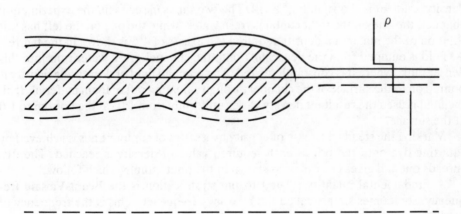

(b)

Fig. 13.10 — Approximations to intrusions between two fluids with a thick interface. (a) A reflected turbulent bore. (b) A reflected undular bore.

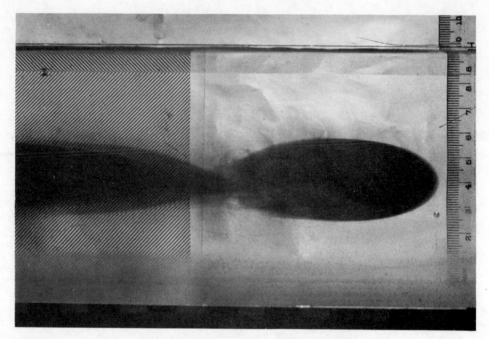

Fig. 13.11 — Undular intrusive bore observed in laboratory experiment, in which the head is almost completely separated from the following flow.

13.2 CONTINUOUS STRATIFICATION

The simple "two-tank" method of filling a laboratory tank with fluid of uniform density gradient is shown in Fig. 13.14. The two tanks in series fill the experimental channel; the one on the right contains fresh water, while the one on the left has salt solution of the maximum density required (an excess of 4% is shown in the figure). As fluid is pumped from the dense tank onto the channel, fresh water runs into this denser tank, where it is constantly stirred. The pumped fluid of decreasing density is poured on to the surface of the channel through a floating raft of foam material. If all the fluid is used the resultant density varies uniformly from 1.04 at the bottom to 1.0 at the surface.

Marking the steady decrease of density by a series of contour lines is achieved by squirting dye onto the raft as each required value of density is reached. The dye spreads out and gives clear lines as shown in the photographs which follow.

A fundamental quantity related to the stratification is the **Brunt-Vaisala** frequency, sometimes simply called the **buoyancy frequency**. This is the frequency of oscillation of a small particle of the fluid if disturbed a small distance up or downwards. It is usually denoted by N (per second), and is given by $-\{g/\rho.d\rho/dz\}^{\frac{1}{2}}$.

In fluid with a constant density gradient, i.e. $d\rho/dz$ is constant, the value of the buoyancy frequency is constant with depth throughout the fluid. In this case an important relationship exists with the Froude number. Using the notation in Fig. 13.15, the density gradient $= -d\rho/dz$, so $\Delta\rho/H = -d\rho/dz$, thus

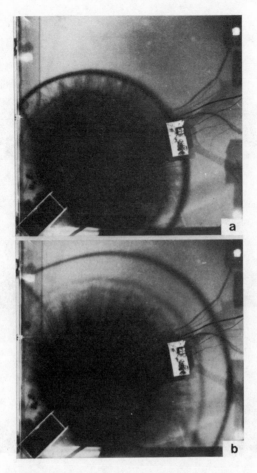

Fig. 13.12 — Sequence of waves generated by an axisymmetric mixed region advancing through a two-layer stratification. (Courtesy of T. Maxworthy).

$$\mathrm{Fr} = \frac{U}{\left(\frac{\Delta\rho}{\rho_0} . gH \right)^{\frac{1}{2}}}$$

And since

$$N = \left[-g \; \rho_0 \; \frac{\mathrm{d}\rho}{\mathrm{d}z} \right]^{\frac{1}{2}}$$

it follows that

$$\mathrm{Fr} = U/NH$$

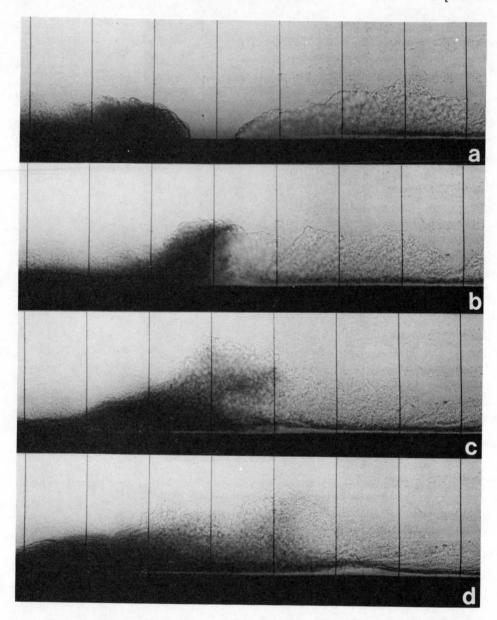

Fig. 13.13 — Four stages in the collision between two unequal gravity currents.

Theory and experiments have shown that the movement of an obstacle through a uniform stratification can generate several different modes of internal waves [9]. The maximum velocity of disturbances which can move upstream of an obstacle of height H is $c = NH/\pi$, or in other words, for disturbances to move upstream the Froude number, U/NH, must be less than $1/\pi$.

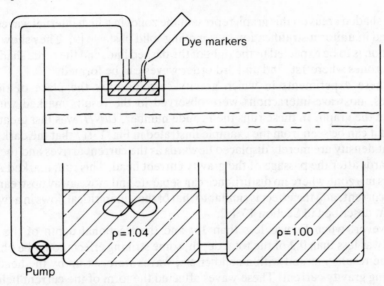

Fig. 13.14 — The "two-tank" method of filling a tank with fluid of uniform density gradient.

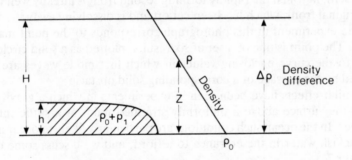

Fig. 13.15 — Notation used for tank with fluid of uniform density gradient.

13.2.1 Gravity currents in uniform stratification

This section will consider the passage of a gravity current along a boundary through a medium which is uniformly stratified, or in other words has a constant density gradient.

The speed of a gravity current travelling through a limited depth of stratified surroundings is close to that of a gravity current in a uniform fluid of the mean density. The important differences are related to the interaction with the internal waves that can be generated by the advance of the gravity current.

In the experiments to be described the buoyancy frequency, N, was varied between 1.3 and $2.0\,\mathrm{s}^{-1}$, and the Froude number U/NH between 0.05 and 0.4. Some of the experimental results for gravity currents of different sizes and values of the Froude number are shown in Fig. 13.16.

The shaded areas on this graph represent the zones in which internal waves can be produced in uniform stratification by a moving solid obstacle [9]. They show that no interaction is to be expected to the right of the dashed line, and the three shaded area enclose zones where 1st, 2nd and 3rd order waves can be formed.

In these experiments in which gravity currents took the place of the solid obstacles, no wave-interactions were observed in the results marked with clear circles in the graph. In these runs the Froude number, U/NH, was just greater than $1/\pi$, and it can be seen from the example ilustrated in Fig. 13.17 that the dark lines of constant density are merely displaced upwards as the current arrives and then move downwards after the passage of the gravity current head. This run, marked A in the graph, is in a zone where no disturbance can separate and move away upstream of the gravity current, and the effect is similar to that of the supercritical flows in a two-layer medium, described in Section 13.1.2.

However, when Fr was less than $1/\pi$ and the fractional depth of the gravity current was less than 0.2 of the total depth, some striking effects could be observed. In all the results marked with a solid circle, waves were set up by the head of the advancing gravity current. These waves affected the form of the current behind the head in a rhythmical manner. The dense fluid in the original head was cut off from the flow, which formed a second head. The process was repeated and later a third new front appeared. The dissolution of each head was followed by the development of a new one to follow it. A photograph of one such experiment in progress is shown in Fig. 13.18, in which the rapidly forming second front is already well formed, behind the original front, which has been cut off and is dissolving away.

The experiment in this photograph corresponds to the point marked B in Fig. 13.16. The point is one of a set of six results, plotted as a solid circle, which are all close to the curve enclosing values in which first mode waves are forecast to be formed by the passage of a corresponding solid obstacle.

Similar effects have been seen in experiments in which a gravity current flows along the surface above a uniformly stratified fluid, at a sufficiently low Froude number. In the ocean such conditions are particularly likely to occur near the surface in brackish water in the entrance to a fjord, and we discuss some observations in Chapter 7.

13.2.2 Intrusions into constant density gradient stratification

Intrusive gravity currents are produced in the environment in surroundings of constant density gradient by the collapse of a region which has become mixed by turbulence; this mixed fluid then moves forward as a gravity current, flowing at a height appropriate to its density. These intrusions may be responsible for the formation of internal waves whose properties have been examined in laboratory experiments.

The complete process of collapse has been divided into three stages. After the initial stage of rapid collapse, the principal stage follows which is also controlled by gravitational forces. In the final stage the flow is controlled by viscous forces and is in the form of a long thin wedge.

Internal waves similar to those formed during the early stage of the collapse can conveniently be generated by an oscillating plunger [10]. The pattern of these

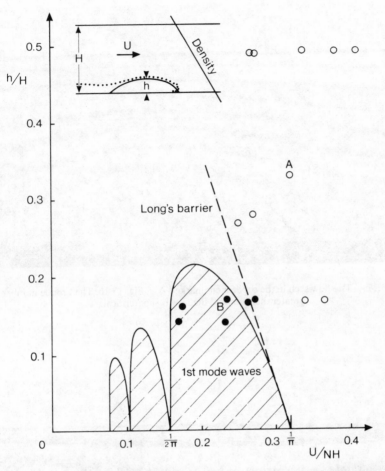

Fig. 13.16 — Experimental results from gravity currents of different size and Froude number, advancing through fluid of uniform density gradient. Shaded circles, ●, show values where interaction was seen between the gravity current and the waves it produced.

internal waves can be represented by moving rays connecting either wave crests or troughs. At first these rays decrease their slopes, but a simple steady-state pattern is formed in the later stages. Fig. 13.19 represents the first two stages and the wave pattern. It has been shown that the angle θ of the rays is given by

$$\theta = +\sin^{-1}(f/N)$$

where f is the frequency of the plunger oscillations and N is the buoyancy frequency.

Examination of the progress of gravity currents moving through uniform stratification, both in two-dimensional and also as three-dimensional radial flows, shows the pulsing flow already described for boundary gravity currents [11]. As noted above, this behaviour is a result of strong interaction between the wave system and

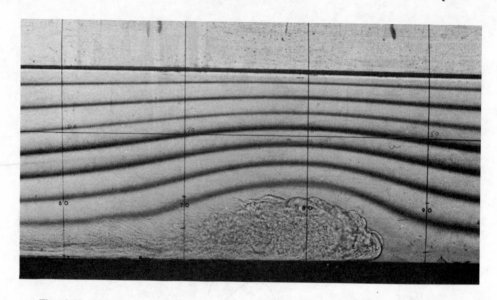

Fig. 13.17 — The flow seen in the experiment marked A in Fig. 13.16. The Froude number is greater than $1/\pi$, and the flow is supercritical.

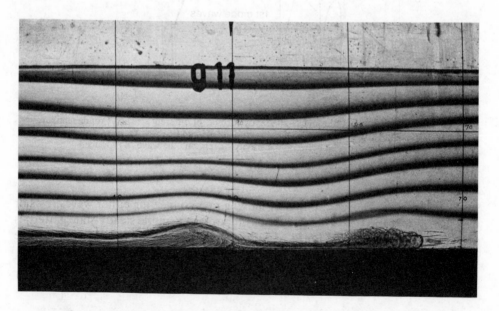

Fig. 13.18 — One stage in the experiment marked B in Fig. 13.16. The front generated first-mode waves in the stratified fluid, which interacted with the gravity current flow. The original front is dissolving away and a second front has started to form.

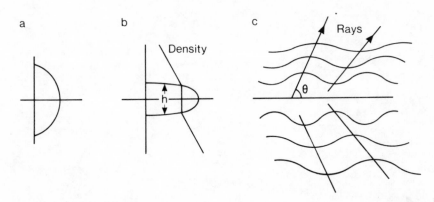

Fig. 13.19 — Stages in the generation of waves by an intrusion into uniformly stratified fluid.

the gravity current in which the current is left behind as it is passed by successive waves.

The conditions in these experiments closely resemble those found both in the atmosphere and in the ocean, which we have already examined in Chapters 4 and 7.

BIBLIOGRAPHY

[1] Baines, P. G. & Davies, P. A. 1980. Laboratory studies of topographic effects in rotating and/or stratified fluids. In: *Orographic Effects in Planetary Flows: Global Atmospheric Research Program Publ. Series*, No. 23, Chapter 8.

[2] Long, R. R. 1970. Blocking effects in flow over obstacles. *Tellus*, 22: 471–480.

[3] Wood, I. R. & Simpson, J. E. 1984. Jumps in layered miscible fluids. *J. Fluid Mech.*, 140: 329–342.

[4] Yih, C.-S. & Guha, C. R. 1955. Hydraulic jump in a fluid system of two layers. *Tellus*, 7: 358–366.

[5] Chu, V. H. & Baddour, R. E. 1977. Surges, waves and mixing in two-layer density stratified flow. *17th Congress, IAHR*, Baden-Baden, vol. 1: pp. 303–310.

[6] Simpson, J. E. & Rottman, J. R. Internal bores in the atmosphere and in the laboratory. To be submitted to *Quart. J. Roy. Met. S.*

[7] Britter, R. E. & Simpson, J. E. 1981. A note on the structure of an intrusive gravity current. *J. Fluid Mech.*, 112: 459–466.

[8] Maxworthy, T. 1980. On the formation of nonlinear internal waves from the gravitational collapse of mixed regions in two and three dimensions. *J. Fluid Mech.*, 96: 47–64.

[9] Long, R. R. 1955. Some aspects of the flow of stratified fluids, III. Continuous density gradients. *Tellus*, 3: 341–357.

[10] Wu, J. 1969. Mixed region collapse with internal wave generation in a density-stratified medium. *J. Fluid Mech.*, 35: 531–544.

[11] Amen, R. & Maxworthy, T. 1980. The gravitational collapse of a mixed region into a linearly stratified fluid. *J. Fluid Mech.*, 96: 65–80.

14

Gravity currents in a turbulent environment

An example of a turbulent flow can be experienced by standing out of doors in a wind. Although the wind strength may be indicated for example as 10 m s^{-1}, rapid variations of several metres per second can be felt. These arrive as gusts then die away and the process is continually repeated. The flow can be pictured as a mixture or a collection of eddies, all of different sizes and strengths, and the speed of 10 m s^{-1} can only represent some kind of averaged value.

When the front of a gravity current moves in calm surroundings, most of the mixing occurs in the region of the head and this results in the laying down of a stable layer above the following current. In cases when the level of ambient turbulence is small it usually has only a small effect on this mixing process, but if the continued spread of a finite amount of dense fluid into a turbulent environment is considered this may no longer be true. Although at the start the normal processes dominate, eventually as the density becomes smaller and the velocity of the front is sufficiently reduced, the effect of external turbulence may become appreciable. When the current has reached this stage the form of the leading edge begins to change and the depth of the current, which had been decreasing, may begin to increase.

14.1 TURBULENCE EXPERIMENTS

Many investigations have used the theoretical concept of turbulence in which the statistical properties do not vary with position and have no preferred direction. This is known as "homogeneous isotropic turbulence".

A typical grid used for generating turbulence in a wind- or water tunnel is shown in Fig. 14.1. Turbulent energy is produced in the flow past the grid; this becomes almost isotropic and homogeneous with distance downstream.

Another way of producing turbulence is by regular oscillations of this kind of grid. This method has been used to generate turbulence in water tanks, especially to examine the effects of turbulence on stratified fluids. A measure of the turbulent intensity produced by the oscillation of such a grid at a specified distance can be given in terms of the values of representative velocity fluctuations U' and length fluctuations L'. Extensive experimental measurements have been made of such turbulent properties produced by grids of different sizes and proportions, oscillating at various frequencies [1,2].

14.1.1 Experiments using oscillating grids

An oscillating grid was used in a series of experiments [3] to maintain and enhance the shear turbulence in a streaming flow over a stationary saline gravity current. As shown in Fig. 14.2, the moving floor passed beneath a fixed-plate working section. The belt travelled at the mean-depth speed, U, of the flow in order to minimise mean velocity shear in the stream. The turbulence generated by an oscillating grid overhead propagated downwards, whilst being advected by the stream. Unrepresentative mixing of the entering current was suppressed by the shield plate.

Salt solution was continuously supplied at the downstream end of the gravity current in order to compensate for fluid transported into the stream. The supply was adjusted until the front of the gravity current stabilised at the leading edge of the fixed floor. Using this apparatus it was possible to maintain turbulent intensities U'/U at any level from zero (the normal gravity current) to about 10 or more, the kind of value found in tidal flows. At these large intensities the frontal slope appeared as a wedge, with a fluctuating well-defined interface separating the undiluted current flow and the turbulent surroundings. The current, maintained by a continuous flow from behind, and protected from turbulence, was eroded, not diffused. Fig. 14.3 shows the non-dimensional mixing rate appropriate for gravity currents in non-turbulent streams plotted against turbulent intensity U'/U. The measurements approach normal gravity current results [4] for $U'/U < 0.5\%$, to the left of line A in the diagram. The mixing flux becomes independent of stream speed for $U'/U > 50\%$, to the right of line B, as shown by the approach to a gradient of 3 in the log–log plot. Here the mixing rate is 2 to 4 times larger than the non-turbulent values.

In these experiments the process seems more akin to erosion rather than turbulent diffusion, and the results were found to be consistent with those obtained [1] in oscillating grid experiments in a static tank.

14.1.2 Dissipation of fronts

Laboratory experiments on the effects of turbulence have been related to the spreading of salinity in estuaries [5]. A long rectangular channel was used with inflow of fresh water at one end and a portion at the opposite end in which a constant salinity was maintained. The mixing effect of the tide was simulated by turbulence generated by oscillating screens at various amplitudes and frequencies.

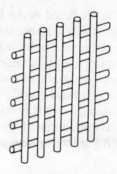

Fig. 14.1 — Typical turbulence generating grid.

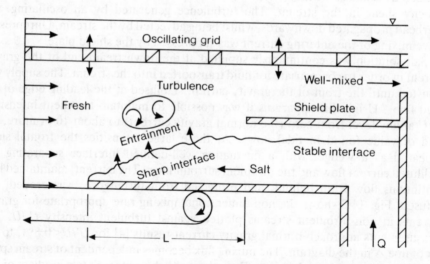

Fig. 14.2 — Apparatus for measuring the effect of turbulence on a gravity current in a steady state.

A turbulent diffusion coefficient K was determined with zero density difference, dye being used as tracer, using a one-dimensional convective diffusion equation. A larger, gross, diffusion coefficient K' is measured in the presence of a density difference-induced gravitational convective circulation. The ratio K'/K was taken as a measure of the degree of mixing or stratification in an estuary.

Rising bubbles have been used as an alternative to grids or screens to generate turbulence in laboratory experiments. In particular the effect on front formation and

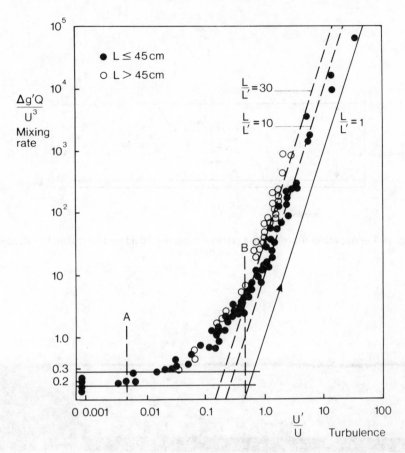

Fig. 14.3 — Graph of the mixing rate in a gravity current subjected to different intensities of turbulence.

dissipation in buoyancy-driven flows has been examined in turbulence produced in this way [6].

Lock-exchange was used to establish a saline gravity current flow, which was kept turbulent by bubbling air from a manifold in the base of the tank, as shown in Fig. 14.4. When the gate is opened a gravity current begins to flow, but mixing by the turbulence starts to transfer momentum between the two counterflowing layers and reduced the vertical stratification. Eventually a vertically mixed state is established in which there is an expanding region of horizontal density gradient, centred on the original position of the partition. Fig. 14.5 illustrates the stages in the change from a gravity current front to such a vertically mixed region.

The advance of the gravity current front is plotted against the elapsed time in non-dimensional form in Fig. 14.6 for a number of different values of g', the original reduced gravity, and H, the depth of the tank. In an undisturbed environment the

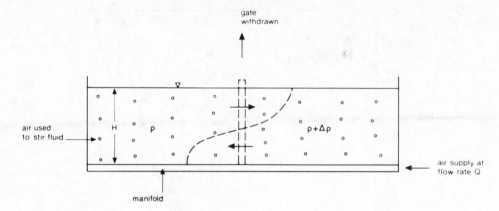

Fig. 14.4 — Apparatus for releasing a gravity current into liquid made turbulent by ascending air bubbles.

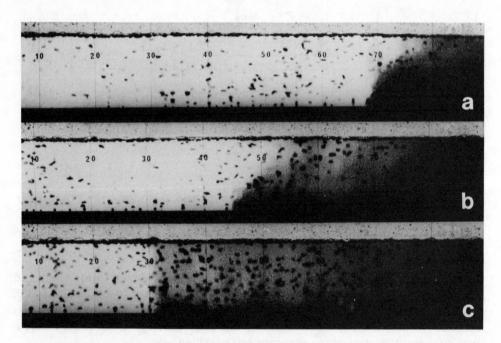

Fig. 14.5 — Three stages in which a gravity current changes into a vertically mixed area in a turbulent water tank.

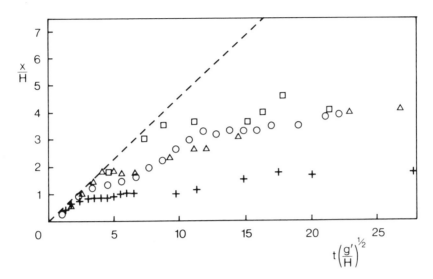

Fig. 14.6 — The advance of the front of a gravity current with time in turbulent surroundings. In an undisturbed environment a gravity current travels at the rate shown by the broken line. Values of g': + 3.1, $\triangle$ 6.7, $\bigcirc$ 12.6, $\square$ 15.6 cm s^{-2}.

current flows at the constant speed shown by the broken line. In the initial stages the data follow this line but the turbulence eventually retards the current. The departure from the line at long times increases with increasing g' and H, and it is possible to determine an eddy "diffusivity" K for the longitudinal transport.

It should be emphasised that in the absence of turbulence the longitudinal density flux is maximal and the presence of the turbulence decreases the flux.

14.1.3   Formation of fronts

A sharp front can be formed from a horizontal density gradient when the turbulence is reduced. This situation occurs in the atmosphere, where the formation of a sea-breeze front may be inhibited until the latter part of the day. As the level of convective turbulence dies down a sharp sea-breeze front may develop.

The formation of a sharp front from a horizontal density gradient was investigated [7], using the apparatus shown schematically in Fig. 14.7(A). In this "tilting-tank" experiment a vertical density gradient with horizontal density contours marked by dye was set up, using the two-tank method, in a vertical tank. The tank is then rapidly rotated to a horizontal position and the resulting flows observed as the density markers return to the horizontal position.

When the density gradient is uniform, as marked by a solid line in Fig. 14.7(B), the lines show a steady return to the horizontal, except for some disturbances near the ends of the tank. However, where the density gradient is steepened in one half (but with no density step), as shown by the dotted line, a steadily sharpening front develops.

Fig. 14.8 shows two stages in the development of one of these experimental fronts

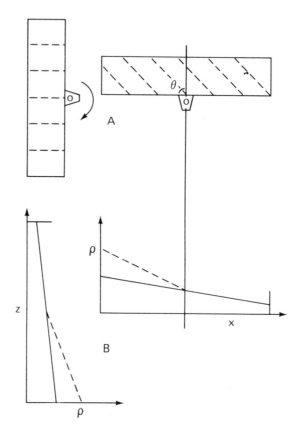

Fig. 14.7 — Tilting tank apparatus used to examine the formation of a gravity current front. (A) Experimental arrangement. (B) Density gradients: —— uniform, – – – – gradient change at centre of tank.

from a non-uniform horizontal density gradient. The converging density lines can be seen in (a) and a sharp shadowgraph line has appeared at the developing front. In (b), as many as four "density lines" have converged at the front, which is now well developed with turbulent mixing at the leading edge.

BIBLIOGRAPHY

[1] Turner, J. S. 1968. The influence of molecular diffusivity on turbulent entrainment across a density interface. *J. Fluid Mech.* 33: 39–56.
[2] Hopfinger, E. J. & Toly, J. A. 1976. Spatially decaying turbulence and its relation to mixing across density interfaces. *J. Fluid Mech.,* 78: 155–175.
[3] Thomas, N. H. & Simpson, J. E. 1985. Mixing of gravity currents in turbulent surroundings: laboratory studies and modelling implications. In: *Turbulence and Diffusion in Stable Environments*, ed. Hunt, J. C. R. Clarendon Press, Oxford, pp. 61–95.

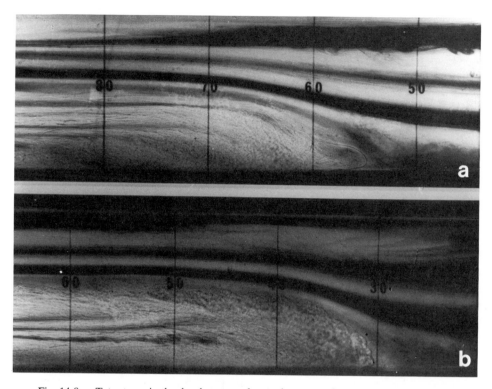

Fig. 14.8 — Two stages in the development of a gravity current front in a non-linear density gradient. In (a) a sharp line appears at the front; in (b) a fully developed turbulent front has formed.

[4] Britter, R. E. & Simpson, J. E. 1978. Experiments on the dynamics of a gravity current head. *J. Fluid Mech.*, 88: 223–240.

[5] Harleman, D. R. F. & Ippen, A. T. 1960. *The turbulent diffusion and convection of saline water in an idealised estuary*. Int. Assoc. Sci. Hydrol. Commission of Surface Waters, publ. no. 51, pp 362–378.

[6] Linden, P. F. & Simpson, J. E. 1986. Gravity-driven flows in a turbulent fluid. *J. Fluid Mech.*, 172: 481–497.

[7] Linden, P. F. & Simpson, J. E. 1987. Frontogenesis in a fluid of non-uniform density. Submitted to *J. Fluid Mech.*

15

Viscous gravity currents

Chapter 11 illustrated the change in form of the head of a gravity current for different values of the Reynolds number. The behaviour of many gravity currents in the environment appears to be independent of viscous effects, since no variation in the properties of the flow is apparent with changes in Reynolds number. For example, in the case of thunderstorm outflows in the atmosphere, although there must be small-scale viscous effects near the ground, it is known that their effect on the general dynamics of the flow is negligible provided the Reynolds number, UL/ν, is greater than a certain value. The magnitude of this critical value has been shown in laboratory experiments on gravity currents [1] to be between 500 and 1000, so thunderstorm outflows, with a Reynolds number of over 10^6, are well beyond this limit.

However, there are many large-scale examples such as lava flows and currents of debris or mud, in which viscous forces play a large part in the dynamics of the flow. Useful insight into the behaviour of this type of viscous flow can also be obtained from laboratory experiments.

Low Reynolds number experiments can be carried out with flows of salt solution in water, provided that both the size and the density differences are kept very small. One of these shallow, slow, currents is pictured in Fig. 15.1; the speed of this flow is about 0.5 cm s^{-1} and the depth is 0.5 cm, so the Reynolds number Uh/ν of the flow is $0.5 \times 0.5/0.01$ or 25. This flow is very smooth and the elevated head of the current is small; for values of Re less than 10 the head no longer exists and the flow is a smooth wedge.

For experiments with much lower Reynolds number a viscous material like treacle has been used, and there are a number of silicone oils available to cover the range of viscosities required.

15.1 INERTIAL AND VISCOUS REGIMES

In some environmental flows, for example in the spread of oil on the surface of the sea, although viscous effects may not be significant at first, they become more important in the later stages when the layer of oil becomes thin and moves more

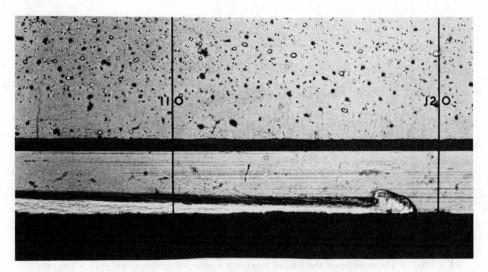

Fig. 15.1 — A saline gravity current with Reynolds number of 25.

slowly. A stage is finally reached when the flow is controlled mainly by viscous forces. (In the case of spreading oil, surface tension forces may also become important in the later stages.)

From dimensional analysis [5] it has been shown that

$$\frac{\text{inertial force}}{\text{viscous force}} = \frac{\rho D^2 h/t^2}{\rho \nu D^2/th} = \frac{(h/t)h}{\nu}$$

which can be written Uh/ν, the form we have already described for the Reynolds number of the flow. So we see that the Reynolds number, Uh/ν, can be expressed as the ratio of the inertial to the viscous forces determining the flow.

Fig. 15.2 shows results of a series of experiments in which a fixed flux of salt water was released into an axisymmetric tank containing much deeper fresh water [2]. The position of the leading edge of the flow, normalised by

$$r_* = (Q_1 g')^{\frac{1}{4}} t^{\frac{3}{4}}$$

Its expected temporal variation is plotted against t/t_*, where the time is normalised with the viscous time scale [3],

$$t_* = Q^{\frac{1}{4}}/(\nu g')^{\frac{1}{4}}$$

The experimental results approximate to a single curve, and demonstrate clearly the

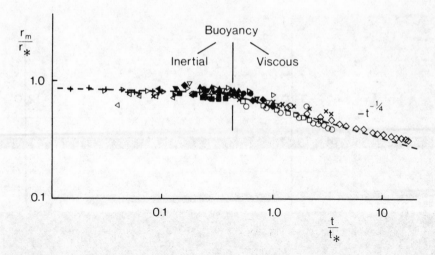

Fig. 15.2 — The position of the leading edge of a number of axisymmetric gravity currents,
showing the passage from the inertial to the viscous regimes. (Courtesy of R. E. Britter).

shift from the buoyancy–inertial regime to the buoyancy–viscous regime as each flow
proceeds.

15.2   THE SPREADING OF VISCOUS GRAVITY CURRENTS

The spread of both two-dimensional and axisymmetric currents over a rigid horizontal surface has been studied theoretically and the results compared with laboratory
experiments [4,5]. Except possibly in the early collapse stages, the length of the
current will greatly exceed its thickness, which allows for the boundary-layer
approximation common to lubrication theory to be used in the analysis. The effect on
the gravity current of the motion of the upper fluid can be entirely taken into account
by applying the boundary condition of zero stress at the upper surface of the current.
The current is thus identical with one propagating with a free surface beneath a fluid
of negligible inertia.

The spread of viscous gravity currents is found to be as follows:

Radial: $r_N = 0.623(g'Q^3/\nu)^{\frac{1}{8}}\, t^{\frac{1}{8}}$

Two-dimensional: $x_N = 0.804(g'q^3/\nu)^{\frac{1}{5}}\, t^{\frac{4}{5}}$

One important result from this theoretical treatment is that conditions at the
front of a viscous current play no part in determining its motion or shape. This lack of
influence of the front will only be true if the Reynolds number is low enough and if
the effects of surface tension are small.

This is different from high Reynolds-number gravity currents which are totally
controlled by conditions at the front. As has been described in this book, many
theoretical and experimental studies have been devoted to determining the controll-

ing condition, or Froude number, at the front when the Reynolds number is very great.

15.3 VISCOUS FLOWS ON SLOPES

If viscous fluid is released on a horizontal surface it takes up a circular plan form which is stable to small disturbances. However, if fluid is released on a sloping surface — for example some liquid detergent on a sloping plate — a very different plan form occurs. If a broad band of viscous fluid, uniform in depth across the slope, is released it first moves steadily downstream and then breaks up into a series of waves of increasing amplitude. An example is shown in Fig. 15.3. It has been shown

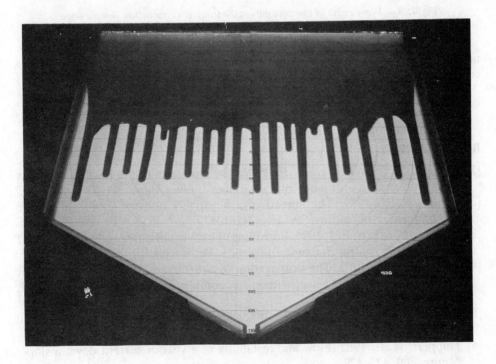

Fig. 15.3 — Instability at the front of a viscous flow down a slope. (Courtesy of H. E. Huppert).

that the instability cannot be explained by viscous effects alone and that surface tension must be taken into account [6]

Experiments using silicone oils indicate that the wavelength of the instability is independent of the coefficient of viscosity. Other experiments, with two fluids of comparable viscosity but different surface tension, show that the wavelength is a function of the surface tension.

The graph in Fig. 15.4 shows a plot of experimental results [6] of the wavelength,

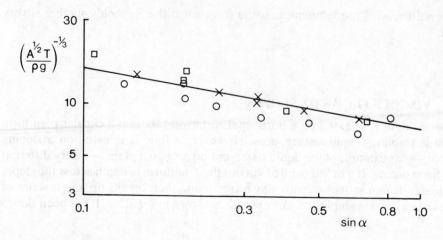

Fig. 15.4 — Wavelength of the instability at the front of a viscous flow down a slope.

normalised with respect to $(A^{\frac{1}{2}}T/\rho g)^{\frac{1}{3}}$, as a function of the slope angle, α. The data collapse onto a single curve and are well represented by

$$\lambda = 7.5 \, (A^{\frac{1}{2}}T/\rho g \sin \alpha)^{\frac{1}{3}}$$

BIBLIOGRAPHY

[1] Simpson, J. E. & Britter, R. E. 1979. The dynamics of the head of a gravity current advancing over a horizontal surface. *J. Fluid Mech.*, 94: 477–485.

[2] Britter, R. E. 1979. The spread of a negatively buoyant plume in a calm environment. *Atmos. Environ.* 13: 1241–1247.

[3] Chen, J.-C. & List, E. J. 1976. Spreading of buoyant discharges. *Proc. 1st CMIT Seminar on Turbulent Buoyant Convection*, Dubrovnik, pp. 171–181.

[4] Huppert, H. E. 1982. The propagation of two–dimensional and axisymmetric viscous gravity currents over a rigid horizontal surface. *J. Fluid Mech.*, 121: 43–58.

[5] Didden, N. & Maxworthy, T. 1982. The viscous spreading of plane and axisymmetric gravity currents. *J. Fluid Mech.*, 121: 27–42.

[6] Huppert, H. E. 1982. Flow and instability of a viscous current down a slope. *Nature*, 300: 427–429.

16

Suspension flows

The density difference in gravity currents can be caused by thermal effects or as the result of dissolved material. Another very important type of gravity current is that in which the density difference is caused by *suspended* material. These suspension gravity currents are of special interest to the geologist, the oceanographer and the civil engineer. There is a wide range of such flows, which include streams of muddy or turbid water, powder-snow avalanches, and flowing material from some kinds of volcanic eruptions.

If the suspended material which causes the increased density of the fluid consists of very fine particles, then the rate of fall-out may be very small. For a considerable time the behaviour of this suspension current will be similar to that of a gravity current caused by dissolved material.

An example of such a suspension current can often be seen in the atmosphere when a large building is being demolished. As the building collapses, air is displaced, together with large amounts of very fine suspended dust particles; this results in a suspension gravity current moving away from the building. A good example is shown in Fig. 16.1, which is a view of one of these gravity currents moving towards the observer. (Part of the building can still be seen standing behind the current, since in this case it only collapsed half its original height.)

16.1 TURBIDITY CURRENTS

Suspension currents are also common in water. They can easily be demonstrated in clear, calm pond water by using a stick to stir the mud at the bottom of a pond. A turbidity current of muddy water will spread along the bottom of the pond away from the disturbance. Similar muddy currents can be seen when a clear stream is joined by a muddy river.

The mechanism of flows of muddy, or turbid, water was first recognised in the study of sediment-laden streams continuing beneath the surface of larger bodies of water. Fig. 16.2 shows a diagram of such a flow of dense turbid water entering a lake or reservoir and starting to flow as a turbidity current beneath the less dense fresh

Fig. 16.1 — Gravity current formed by suspended dust from a partly demolished building.
(Photo by Associated Press).

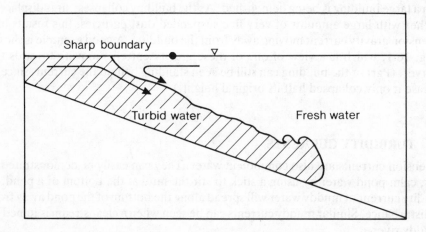

Fig. 16.2 — Turbidity current flowing into a large area of fresh water.

water. Where a turbid river enters a lake a sharp boundary between clear and muddy water is visible at the surface.

It is known that a turbidity current on the sea bed can erode a deep channel, rather as a flowing river forms a channel. The sediment transported and deposited by turbidity underflows in this way is of great significance in the development of the sea bed and in the formation of sedimentary rocks.

16.1.1 Experiments on turbidity currents

The creation of a controlled laboratory turbidity current involves considerable practical difficulties, in contrast to the relative ease with which saline currents can be created and their density measured.

The first experiments on turbidity currents were described in 1951 [1], and showed that high-density turbidity currents could be produced in the laboratory. A dense mixture of sand, silt and water was released down a short slope under clear, still water. Although the vertical distance of fall was only about 50 cm, this was sufficient to give an underflow moving at more than 5 cm s^{-1} which then continued along the full length of about 3 m of a horizontal channel. A front view of such a laboratory turbidity current has already been shown in the first chapter of this book, in Fig. 1.3.

From such experiments it was concluded that oceanic turbidity currents falling through a distance of several kilometres could reach the high velocities deduced from deep sea evidence.

Except in cases where the deposition of the suspended solid material is negligible during the time of the experiment, the simple lock-exchange technique used for saline gravity currents is inadequate. In later experiments mechanisms were devised for introducing the material to be suspended to start a turbidity current in "running order" [2,3]. Fig. 16.3 shows one such arrangement. The material to be suspended is submerged in the introduction box and when the gate is removed the material drops and forms a suspension current. This current forms an underflow from which material settles out on the bed of the tank.

Such turbidity currents show most of the features already observed in saline gravity currents [4], for example in the nature of mixing at the head and the relationships between velocity, depth and density difference, both along level floors and down slopes. One notable difference, however, was the threshold for Re-independent flows, beyond which the viscous regime was entered. It was found that this regime was first attained in these flows at appreciably higher Reynolds numbers [3].

These experiments have led to some understanding of the transport and deposition of the sediments carried by turbidity currents. In one typical series the material used for the sediment was a moderately well sorted mixture of plastic beads of density 1.52 g cm^{-3} and mean diameter 0.18 mm. In these experiments the mode of deposition from the travelling bead suspension and the character of the vertical grading of the deposit were studied.

Study of the processes of erosion and deposition gave a picture as shown in Fig. 16.4 in which for a short distance behind the head the erosion process dominates, and then follows an area of increasing deposition.

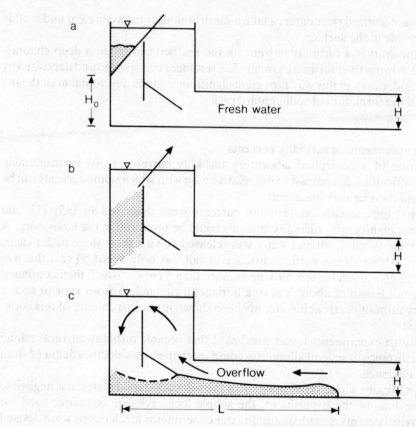

Fig. 16.3 — Experimental arrangement for production of turbidity currents. (a) Material submerged in introduction tank. (b) Gate removed. (c) Underflowing turbidity current. (After Riddell, 1969 [3]).

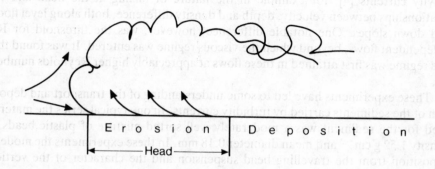

Fig. 16.4 — Areas of erosion and deposition at the front of a turbidity current.

16.1.2 High density turbidity currents

Details of the internal flow of a suspension current as shown in Fig. 16.4 are usually
not visible from outside, but have been deduced in field work by the use of pressure
sensors on the ground beneath the flow. Sharp separation of the fluid into two (or
more) regimes has been established.

Laboratory experiments on high density turbidity currents have been used to
examine such two-phase suspension currents. Fig. 16.5 shows one such experiment in

Fig. 16.5 — Laboratory two-phase turbidity current. (Courtesy of A. Taira).

progress [5], with the low density Newtonian flow above, and the high density
"body" of non-Newtonian flow beneath it. The initial composition of the fluid used in
this experiment was Bentonite (8%), sand (38%) and water (54%), forming a slurry
of relative density 1.4.

16.2 AIR SUSPENSION CURRENTS: FLUIDISATION

Some of the most violent gravity currents occurring in nature depend on suspension
of solid material by air or other gases. Examples are avalanches of airborne snow,
and pyroclastic flow eruptions from volcanoes, both of which may travel at speeds of
over 100 miles per hour.

16.2.1 Fluidisation experiments

An important process in suspension gravity currents is "fluidisation", whereby the particles are suspended in the fluid, which acquires the macroscopic properties of the combination. A favourite demonstration of the process is illustrated in Fig. 16.6 in

Fig. 16.6 — Plastic duck floating in a fluidised material.

which a toy plastic duck is placed at the bottom of a container with a layer of uniform dense particles. When a stream of air is passed through holes in the base of the tank, the particles are separated from their neighbours by the air rising between them and the contents of the tank begin to move around violently and look like a boiling fluid. At this point the buoyant duck floats to the surface. When this experiment is carried out in a long tilted tank, the fluidised contents begin to flow down the slope, thus forming a kind of gravity current in which solid particles are suspended in air.

The apparatus used in some experiments [6] with a gaseous fluidising medium is shown schematically in Fig. 16.7. A high pressure air supply is regulated by valves

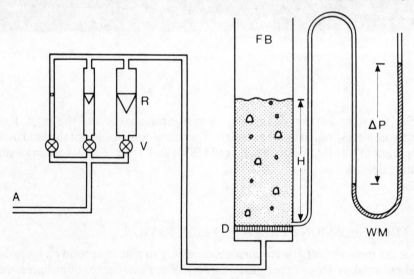

Fig. 16.7 — Schematic diagram of apparatus for fluidisation measurements. (After Wilson, 1980 [6]).

through rotameters and the distributor into the fluidised bed. During runs, the height of the bed is recorded, together with the pressure drop on a water manometer.

Using small smooth glass spheres to permit velocity measurements, one can derive a fluidisation plot (Fig. 16.8) showing two straight lines. These intersect at,

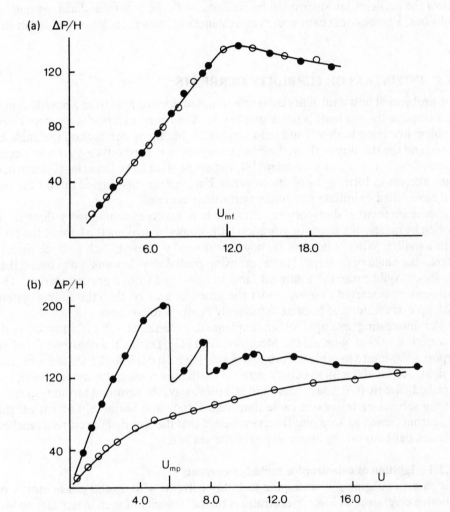

Fig. 16.8 — Fluidisation plots. (a) Glass spheres. (b) Quartz sand. ● Increasing, ○ Decreasing U.

and define, U_{mf}, the speed where the drag force equals the weight. There is commonly some degree of **hysteresis** evident, i.e. a difference between measured curves for increasing and decreasing U. The hysteresis is related to the existence of a wide grain size distribution and also to an irregular particle shape. Graph (a) in Fig. 16.8, using uniform spheres, shows no hysteresis. Quartz sand has both a wide grain size variation and an irregular shape, giving a plot, graph (b), with a large hysteresis.

The grain size variation is so large that the material can never be fully fluidised in the strict sense; before the larger particles are supported, the smallest particles being lost.

The concept of fluidisation by the incorporation of a gas or a liquid can be extended to include the idea of an "auto-suspension" of moving particles [7]. In such a flow the particles are suspended by collisions with the ground and also by mutual collisions, a process that can occur in an avalanche of snow particles suspended in air.

16.3   INITIATION OF TURBIDITY CURRENTS

The problem of how a turbidity current is initiated, for example from a muddy slump on a slope at the sea bed, is not a simple one. The transformation is a complicated problem involving both soil and fluid mechanics. Slumping depends on the angle of slope, and the stability of the sediments; it may occur in earthquakes, even on slopes of only 2 or 3° if shocks are sufficient [8]. Experiments in Lake Superior [9] have had some success in forming turbidity currents, but experiments carried out on the sea bed have failed to initiate any runaway turbidity currents.

Some successful laboratory experiments have involved setting the sediments in motion by means of a shock or vibration [10], simply by striking the base of the tank with a mallet. When a heap of homogeneous sand was used, such a shock merely altered the angle of its slope. However, using graded clay deposits it was found that the shock could cause the sediment mass to move and form a gravity current. The sediment mass started to move under the action of gravity, then the sliding system took up extra water and became sufficiently fluidised to continue.

An interesting example of an unplanned experiment on turbidity currents occurred in 1979 at Nice, on the Mediterranean [11]. During the construction of an airport a 300-metre long earthwork that was being built out into the sea slid away out of sight. About 400 million cubic metres of earth, and some earth-moving vehicles, vanished. Nearly four hours later, and 60 miles away, the resulting turbidity current cut the submarine telephone cable that joins Genoa and Majorca. Later it cut the cable from Genoa to Sardinia. It was estimated that the speed of the current reached 25 miles per hour on the upper slopes of the sea bed.

16.3.1   Ignition of catastrophic turbidity currents

It is easy to imagine self-sustaining turbidity currents with a solid phase mostly of cohesive clay, since at low concentrations the fall velocity of such material is so low that it is only very slowly deposited. However, there is evidence from submarine canyons that highly erosive turbidity currents occur where the only available material is medium-to-coarse silts and sands. A proper formulation of bed sediment entrainment is therefore essential for a clear understanding of an eroding and depositing turbidity current.

If U is the mean down-channel flow velocity, S is the bottom slope, and v is the sediment fall velocity, it is known [12] that a necessary condition for a self-sustaining "auto-suspension" current to exist is that

$$US/v > 1$$

It can be shown that a set of critical velocities and concentrations exists below which the turbidity current dies and above which it grows. Above the critical state the turbidity current is said to "ignite", that is it accelerates and entrains sediment, attaining the catastrophic state. Estimates of this catastrophic state suggest that it is highly erosive and capable of scouring out the submarine canyons which have been observed. This process of "ignition" can take place in suspension flows both in water and in air, for example in the formation of avalanches of air-borne snow particles.

16.4 SUMMARY OF SUSPENSION PROCESSES

In Fig. 16.9 are sketched five different processes which can be involved in the maintenance of a suspension gravity current.

Fig. 16.9(a) shows a turbulent suspension of particles with low fall-speeds, forming a gravity current whose behaviour is very similar to that of a saline gravity current.

In Fig. 16.9(b) the particles have a larger fall-speed, and may bounce up from the ground, eroding further particles in the process. Both (a) and (b) can be expected in both liquids and gases.

The remaining three processes are all examples in which fluidisation plays a part. In Fig. 16.9(c) the fluidising material (either liquid or gas) enters the head of the gravity current from the surroundings. The ambient fluid is entrained in a series of Kelvin–Helmholtz billows above the head, but there is also a comparable flow of light fluid ingested beneath the foremost point of the nose. This can provide a supply of buoyant fluid to fluidise the interior of the head.

In the other two fluidisation processes, which have been dealt with in more detail in Chapter 10, the fluidising material (a gas) involved is supplied in different ways. In Fig. 16.9(d) red-hot material from a volcanic eruption releases water vapour from the surface over which it moves, by heating surface water or ground water, or fluids contained in plants. This fluidisation process appears to be an important one in nature. In the final sketch, Fig. 16.9(e), the fluidising gas is produced from the solid particles, either by fracture or by chemical reaction.

In any particular example of a suspension current in nature, it is likely that more than one of these processes will be operating.

BIBLIOGRAPHY

[1] Kuenen, Ph. H. 1951. Turbidity currents of high density. *18th Int. Geol. Congr.*, London. Reports Pt.8, p. 44.

[2] Middleton, G. V. 1966. Experiments on density and turbidity currents. I. Motion of the head. *Can. J. Earth Sci.*, 3: 523–546.

[3] Riddell, J. F. 1969. A laboratory study of suspension-effect density currents. *Can. J. Earth Sci.*, 6: 231–246.

[4] Allen, J. R. L. 1971. Mixing at turbidity current heads, and its geological implications. *J. Sediment. Petrol.*, 41: 97–113.

[5] Taira, A. 1985. Rheology and flow processes of sediment gravity flows. *Earth Monthly*, 5: 391–397. (In Japanese.)

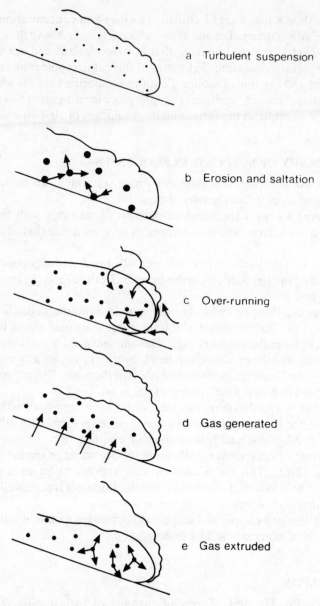

Fig. 16.9 — Summary of processes in suspension currents.

[6] Wilson, C. J. N. 1980. The role of fluidization in the emplacement of pyroclastic flows: an experimental approach. *J. Volcanology & Geothermal Research*, 8: 231–249.

[7] Bagnold, R. A. 1962. Auto-suspension of transported sediment; turbidity currents. *Proc. Roy. Soc., London, Ser. A*, 265: 315–319.

[8] Morgenstern, N. R. 1967. Submarine slumping and the initiation of turbidity

currents. In: *Marine Geotechnique,* ed. Richards, A. F., University of Illinois Press, Urbana, Ill, pp. 189–220.

 [9] Normark, W. R. & Dickson, F. H. 1976. Man-made turbidity currents in Lake Superior. *Sedimentology,* 23: 815–831.

[10] Van der Knapp, W. & Eijpe, R. 1968. Some experiments on the genesis of turbidity currents. *Sedimentology,* 11: 115–124.

[11] Gennesseaux, M., Mauffret, A. & Pautot, G. 1980. Les glissements sous-marins de la pente continentale niçoise et la rupture de cables en mer Ligure. *C. R. Acad. Sci. Paris, Ser D,* 290: 959–962.

[12] Parker, G. 1982. Conditions for the ignition of catastrophically erosive turbidity currents. *Marine Geology,* 46: 307–327.

17

Gravity currents on a rotating earth

Which way does the bath-water rotate as it runs out? Many people have been intrigued by questions about the effect of the Earth's rotation on outflowing bath-water and have looked to see whether it rotates anticlockwise in the northern hemisphere as theory suggests.

In the bath-water example and in many gravity currents the effect of the Earth's rotation is very small indeed and can be neglected. No allowance for it was necessary in the experiments described so far. Nevertheless, in many large-scale environmental gravity currents the Earth's rotation exerts an important influence. Examples in coastal flows from rivers which hug the coast were described in Chapter 7, as were some bottom flows in the ocean moving alongside submerged ridges. Corresponding effects appearing in the atmosphere when large-scale flows meet mountain barriers were described in Section 5.1.

17.1 CORIOLIS FORCE

Because all our measurements are made on a rotating Earth, the path of an object which a stationary observer in space sees as a straight line will appear curved to us on the rotating Earth. We can illustrate the apparent force which causes this acceleration by drawing a straight line radially across a disc rotating in an anti-clockwise direction, as in Fig. 17.1. The pencil follows a straight line in relation to a fixed frame of reference, such as a box containing the disc, but when viewed relative to coordinates moving with the disc, the point swings to the right of its initial line of motion. In the analogous case of the Earth, using our usual rotating reference frame of latitude and longitude, there is a deflection of moving objects to the right in the northern hemisphere and to the left in the southern hemisphere.

The deflecting force is known as the **Coriolis force** and is proportional to the speed of the body and the angular velocity of the Earth (close to 2π radians in 24 hours). The force, per unit mass, also depends on the latitude and can be written as

$$\text{Coriolis force} = fV$$

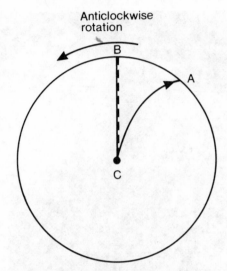

Fig. 17.1—Demonstration of Coriolis force, using a rotating disc. The pencil, which moves in a straight line relative to the box containing the disc, moves in a curve to the right relative to the disc itself.

where f is used to mean $2\Omega \sin \phi$ and is known as the Coriolis parameter. Ω is the angular velocity of the earth $(7.29 \times 10^{-5}\ \text{rad/s}^{-1})$, V is the velocity of the moving object, and ϕ is the latitude.

The relevant dimensionless number which shows the significance of rotational effects is the ratio of inertial to rotational forces, called the Rossby Number. This can be written as $Ro = U/fL$, where U is a velocity, L is a distance and f is the Coriolis parameter. The effect of the earth's rotation can usually be neglected for values of Ro very much greater than 1, but becomes significant when the value of Ro is about unity. For values of Ro much less than 1 the effect of rotation becomes dominant in the flow.

The structure and dynamics of gravity currents in rotating systems are found to be much more complicated than those of density-driven flows without background rotation. In the absence of boundaries, a flow spreading under gravity will approach a state of equilibrium when no further spread is possible. With boundaries present, the Coriolis forces hold the flow against vertical or inclined boundaries and lead to three-dimensional boundary currents.

17.2 SPREADING UNDER GRAVITY : NO BOUNDARIES

When a fluid spreads under gravity in a rotating system, motions normal to the rotation axis induce Coriolis forces which tend to oppose the spreading. In the absence of boundaries or of instability or viscous dissipation, the flow approaches a state of equilibrium. This state is reached when buoyancy and Coriolis forces are in balance, and further release of potential energy is impossible.

Such gravity-driven flows have been examined in a water tank mounted on a

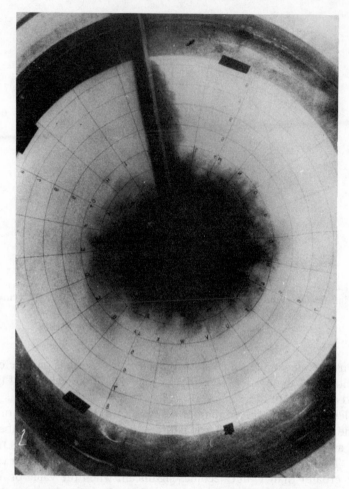

Fig. 17.2 — View from above of instability at the edge of a gravity current spreading in rotating flow. (Courtesy of P. F. Linden.)

rotating table. Various types of instability have been observed at the edge of the dense flow after outward movement has ceased, and Fig. 17.2 shows an example in which a series of large eddies can be seen, rolling up along the leading edge [1].

17.3 SPREADING ALONG BOUNDARIES

Unsteady gravity currents may be produced in the laboratory by the "dam-break" or lock-exchange method as used in two-dimensional gravity currents in non-rotating channels: see Fig. 17.3.

Within an inertial period $2f^{-1}$ (f is the Coriolis parameter), the Coriolis forces become comparable to the buoyancy forces and there is no motion in the direction along the current except along the walls. The "head" and the trailing flow exist much as in the non-rotational gravity current, and the mixing and billows are of the order of

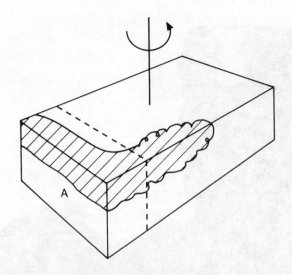

Fig. 17.3 — The spread of a buoyant gravity current after release from a lock at the end A of a rotating laboratory channel.

the current width. There are also rotation-dominated instabilities which have much longer scales.

Power laws to describe the flow do not work, but the nose velocity, depth and width, all scaled by initial conditions, decrease exponentially with time. They scale on the cross-stream length scale $(gH)^{\frac{1}{2}}/f$, where H is the fluid depth. This length scale is referred to as the **Rossby radius of deformation**.

The nose velocity at the wall scaled by local values of $(g'H)^{\frac{1}{2}}$ is constant, as expected from non-rotational theory, and the flow is independent of Reynolds number, UH/ν, for values of Re > 1000. Fig. 17.4 shows the flow of a buoyant current along the right-hand wall after release from a lock in a laboratory channel with anti-clockwise rotation [2].

The most satisfying and simple explanation of the exponential decay is that the rotational systems studied are analogous to intrusions of gravity currents at the bottom of uniform stratification in a non-rotating fluid. Here, internal waves radiate energy away from the current; in the rotating systems, inertial waves can take energy and momentum from the current and disperse it throughout the homogeneous lower layer. Direct measurement of the resulting drag on a solid body has been made and has been referred to as "Coriolis drag".

BIBLIOGRAPHY

[1] Wadhams, P., Gill, A. E. & Linden, P. F. 1979. Transects by submarine of the East Greenland Polar Front. *Deep-Sea Research,* 26A: 1311–1327.

[2] Griffiths, R. W. & Hopfinger, E. J. 1983. Gravity currents moving along a lateral boundary in a rotating fluid. *J. Fluid Mech.,* 134: 357–399.

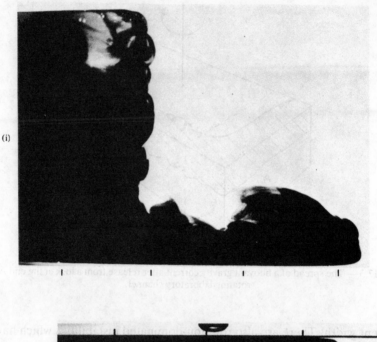

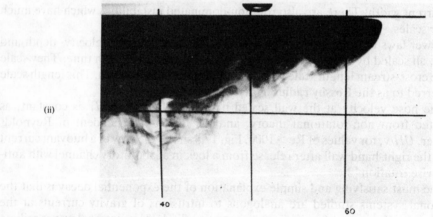

Fig. 17.4 — (i) Top view and (ii) side view of a buoyant gravity current after release in a rotating channel in the conditions shown in Fig. 17.3. (Courtesy of R. W. Griffiths).

18

Numerical models of gravity currents

As previous chapters make clear, laboratory experiments can elucidate the dynamics of large scale environmental gravity currents. Another powerful tool to understanding these phenomena is the numerical model. Using a suitably formulated numerical model it is possible to examine separately a whole range of parameters. For example, in thunderstorm outflows the model can examine the full range of atmospheric parameters (such as wind, temperature stratification, and moisture) which might affect outflow behaviour. In addition, the model can be used to reproduce the basic results of laboratory studies in order to identify any differences between laboratory and environmental flows.

Several different models have been developed to examine dense outflows, and some of them will be described.

18.1   MARKER AND CELL TECHNIQUE

One early example is the marker-and-cell numerical technique used [1] in 1968 to study a two-dimensional gravity current surge in the transient stage after the start from rest. A solute transport equation was coupled with a **Boussinesq approximation** to the Navier–Stokes equations. (In the Boussinesq approximation the difference between the two fluids is neglected in momentum equations, and only appears in density relationships.) The transport equation represented the turbulent diffusion which is known to occur along the shear interface between the gravity current and the environment. In this effectively one-fluid model, the use of this eddy diffusivity in the solute transport equation represents the mixing quite satisfactorily.

Fig. 18.1 shows a series of marker particle plots in the development of a simulated gravity current. The grid mesh is 60 by 20 points, and the ratio of the densities is 1.2. The plots on the left are from a full two-fluid model, and the right series use the Boussinesq approximation in a "one-fluid" solute transport model. The results are quite similar, except for a slightly blunter nose in the one-fluid simulation.

A comparison of these results with measurements of laboratory gravity currents has been made and the results are in good agreement.

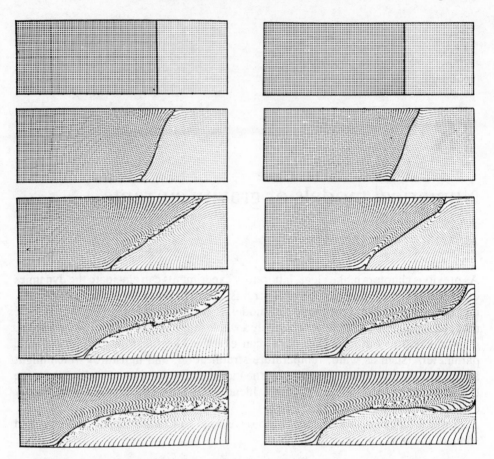

Fig. 18.1 — A two-dimensional gravity current surge modelled by marker particle plots. The left series was produced by a full two-fluid model, and the right series by a Boussinesq approximation to multi-fluid flow. (From Daly & Pracht, 1968 [1]).

18.2 NUMERICAL MODELS OF THUNDERSTORM OUTFLOWS

Some numerical models were specifically designed during the 1970's to examine the flow of the cold dense air which descends from a thunderstorm and spreads out horizontally as a gravity current.

One such model [2] was formed with a time-dependent heat sink designed to represent the evaporative cooling associated with thunderstorm downdraughts. The lateral boundaries were rigid walls, so no environmental flow could be incorporated, and surface friction was introduced through a drag term in the lowest grid level. Fig. 18.2 shows a vertical cross-section of the potential temperature field 12 minutes after the start. The cold air outflow is advancing from left to right, away from the parameterised downdraught, and a clear head structure is evident.

The limitations of this work, for example rigid boundaries and a small computational domain, were largely a consequence of the limited computer resources available at the time. Their study dealt with many questions concerning outflow

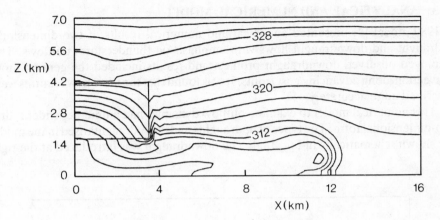

Fig. 18.2 — Potential temperature shown in a two-dimensional model, using a time-dependent heat sink. (From Mitchell & Hovermale, 1977 [2]).

dynamics, for example the sensitivity to ambient stability, surface friction, and the density difference across the gust front, and laid the foundation for future outflow modelling studies.

18.2.1 A three-dimensional outflow model
At about the same time as the work described above, an attempt was made [3] to model the geometry of outflows with an axisymmetric outflow model. This also included modelling of the turbulence structure of the flow. Being axisymmetric, ambient wind shear could not be included in the model.

Fig. 18.3 shows the time-history of the temperature, illustrating clearly the

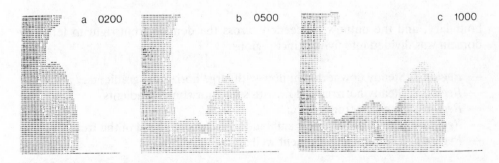

Fig. 18.3 — An axisymmetric outflow simulation, showing cross-section of development of temperature field. (From Teske & Lewellen, 1977 [3]).

development of the outflow and its associated head as the downdraught strikes the ground. There is a wavy structure behind the head along the top of the cold air current moving into the domain initially at rest.

18.3   ANALYTICAL AND NUMERICAL MODEL

A 1980 model [4] presented analytical and numerical results of two-dimensional updraught and downdraught flows with emphasis on thunderstorm outflows. This employed observed downdraught profiles, and results included the generation of convection along advancing gust fronts. It was found that the basic flow features were described quite accurately.

This numerical model to examine outflow dynamics had a time-dependent, dry, two-dimensional form. The types of updraughts and downdraughts used in the model are shown schematically in Fig. 18.4. The downdraught was introduced at the right

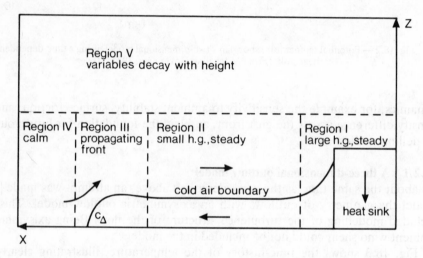

Fig. 18.4 — A schematic diagram showing the various regions of flow identified in the outflow modelling by Thorpe *et al.*, 1980 [4]).

boundary, and the outflow proceeded across the domain from right to left. The domain was divided into five distinct regions:

— *Region 1*. Steady downdraught flow with large horizontal gradients.
— *Region 2*. Steady horizontal flow with small horizontal gradients.
— *Region 3*. The gust front region.
— *Region 4*. The relatively calm ambient environment ahead of the front.
— *Region 5*. The calm environment well above the outflow.

The model domain had dimensions of 80 km horizontal distance by 720 mb vertical pressure range, with grid spacing 500 m by 40 mb respectively; however, the model lateral boundaries allowed gravity waves and the gravity current head to pass out of the domain.

A simple eddy viscosity was incorporated, with no surface friction. The model was run with a uniform ambient wind, but with no vertical wind shear. The time step of the numerical scheme was 15 seconds.

The main drawback to the results of this work was the poor resolution of the numerical model. Observations and laboratory experiments have shown that gravity currents are highly turbulent phenomena, and the numerical model could only represent the bulk flow of such a current. It appeared that higher spatial resolution was needed if numerical models were to resolve the turbulent structure of gravity currents.

18.3.1 Gravity currents and bores

Section 13.1.2 described the production of undular bores in the laboratory using a saline gravity current in a two-layer tank. In a two-dimensional model to study this phenomenon [5], a gravity current was initiated at the left boundary, and while it was moving across the domain from left to right, a low level stable layer was created in the right half of the domain by applying a cooling function near the ground. When the gravity current interacted with the dense layer, it triggered an undular bore which propagated across the domain, as shown in Fig. 18.5.

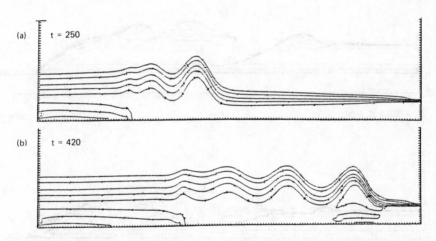

Fig. 18.5 — Potential temperature contours of a numerically simulated undular internal bore.
(From Crook & Miller, 1985 [5]).

18.4 HIGH-RESOLUTION MODELS

Laboratory experiments show that gravity currents usually contain shearing instabilities, but none of the numerical models described so far can reproduce them. The grid resolution is too coarse to resolve the physical turbulence known to be present at gravity current heads.

A two-dimensional numerical model capable of investigating billows at a gravity current front has been developed [6]. The philosophy of this model was to avoid making any approximations to the governing equations. Instead of parameterising turbulence, very high spatial resolution was used to resolve the important small-scale features. A weak background smoothing term was applied to discourage the growth of spurious computational instabilities. The lower boundary was a rigid, free-slip plate.

The model domain was 30 km long and 10 km high with a grid spacing of 100 m in each direction. The flow was initiated as a purely horizontal flow by placing a 2 km high column of cold fluid, with density 2% greater than the environment, at the left boundary, and this was held fixed throughout the simulation.

Initially the outflow was a laminar current with a speed of 18 m s^{-1}. By 10 minutes, wave-like perturbations had developed in the region of sharp density contrast at the top of the outflow. In time these perturbations rolled up into horizontal vortices characteristic of Kelvin–Helmholtz billows. Fig. 18.6(a), shows

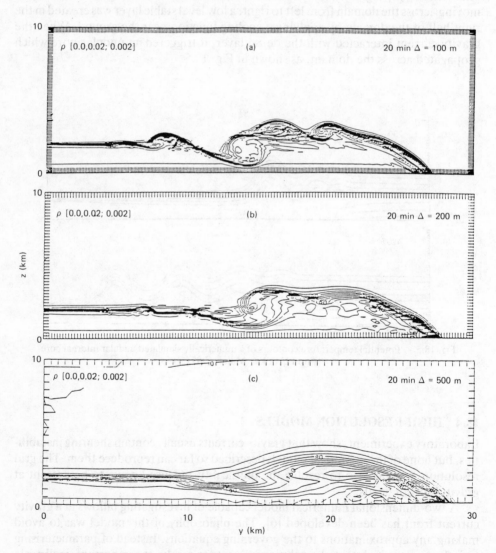

Fig. 18.6 — Three density contour plots of the simulation of a gravity current head. Grid spacings are (a) 100 m, (b) 200 m, (c) 500 m. (From Droegemeier, 1985 [6]).

the density field after 20 minutes, using the 100 m grid spacing. To illustrate the effect of changing the grid spacing, figure (b) shows an identical simulation, but with a 200 m grid spacing, and (c) an identical simulation with a 500 m spacing. It is clear that a grid spacing as fine as 100 m is necessary to model the development of the billows, which have characteristics close to those described in Chapter 11.

The reasons why earlier models did not show evidence of Kelvin–Helmholtz instability include both the lack of spatial resolution and the use of too much numerical smoothing. It seems that the instability along the top of the outflow is very sensitive to the magnitude of the computational mixing.

The time of writing is one of rapid development of numerical models, and we can look forward to fine resolution models in three dimensions. These will study the full development of Kelvin–Helmholtz billows and perhaps also the instability shown in the lobes and clefts.

BIBLIOGRAPHY

[1] Daly, B. J. & Pracht, W. E. 1968. Numerical study of density current surges. *Phys. Fluids,* 11: 15–30.
[2] Mitchell, K. E. & Hovermale, J. B. 1977. A numerical investigation of a severe thunderstorm gust front. *Mon. Wea. Rev.* 105: 657–675.
[3] Teske, M. E. & Lewellen, W. S. 1977. Turbulent transport model of a thunderstorm gust front. *10th Conf. Severe Local Storms,* Omaha, Nebr., pp. 143–149.
[4] Thorpe, A. J., Miller, M. J. & Moncrieff, M. W. 1980. Dynamical models of two-dimensional downdraughts. *Quart. J. R. Met. S.,* 106: 463–484.
[5] Crook, N. A. & Miller, M. J. 1985. A numerical and analytical study of atmospheric undular bores. *Quart. J. R. Met. S.,* 111: 225–242.
[6] Droegemeier, K. K. & Wilhelmson, R. B. 1985. Kelvin–Helmholtz instability in a numerically simulated thunderstorm outflow. *14th Conf. Severe Local Storms,* Indianapolis, Ind., pp. 151–154.

Index